AF559702

Basic Geotechnical Earthquake Engineering

Basic Geotechnical Earthquake Engineering

SECOND EDITION

Dr. Kamalesh Kumar
B Tech., MS, PhD
Assistant Professor
Birla Institute of Technology & Science (BITS)
Pilani, Rajasthan

NEW AGE INTERNATIONAL (P) LIMITED, PUBLISHERS
LONDON • NEW DELHI • NAIROBI
Bangalore • Chennai • Cochin • Guwahati • Hyderabad • Kolkata • Lucknow • Mumbai
Visit us at www.newagepublishers.com

Published by New Age International (P) Ltd., Publishers
First Edition: 2008
Second Edition: 2017

GLOBAL OFFICES

- **New Delhi** **NEW AGE INTERNATIONAL (P) LIMITED, PUBLISHERS**
  7/30 A, Daryaganj, New Delhi-110002, (INDIA)
  Tel.: (011) 23253771, 23253472, **Telefax:** 23267437, 43551305
  E-mail: contactus@newagepublishers.com • Visit us at www.newagepublishers.com
- **London** **NEW AGE INTERNATIONAL (UK) LTD.**
  27 Old Gloucester Street, London, WC1N 3AX, UK
  E-mail: info@newacademicscience.co.uk • Visit us at www.newacademicscience.co.uk
- **Nairobi** **NEW AGE GOLDEN (EAST AFRICA) LTD.**
  Ground Floor, Westlands Arcade, Chiromo Road (Next to Naivas Supermarket)
  Westlands, Nairobi, KENYA, **Tel.:** 00-254-713848772, 00-254-725700286
  E-mail: kenya@newagepublishers.com

BRANCHES

- **Bangalore** 37/10, 8th Cross (Near Hanuman Temple), Azad Nagar, Chamarajpet, Bangalore- 560 018
  Tel.: (080) 26756823, **Telefax:** 26756820, **E-mail: bangalore@newagepublishers.com**
- **Chennai** 26, Damodaran Street, T. Nagar, Chennai-600 017, **Tel.:** (044) 24353401, **Telefax:** 24351463
  E-mail: chennai@newagepublishers.com
- **Cochin** CC-39/1016, Carrier Station Road, Ernakulam South, Cochin-682 016, **Tel.:** (0484) 2377303, **Telefax:** 4051304
  E-mail: cochin@newagepublishers.com
- **Guwahati** Hemsen Complex, Mohd. Shah Road, Paltan Bazar, Near Starline Hotel, Guwahati-781 008,
  Tel.: (0361) 2513881, **Telefax:** 2543669, **E-mail: guwahati@newagepublishers.com**
- **Hyderabad** 105, 1st Floor, Madhiray Kaveri Tower, 3-2-19, Azam Jahi Road, Near Kumar Theater, Nimboliadda
  Kachiguda, Hyderabad-500 027, **Tel.:** (040) 24652456, **Telefax:** 24652457
  E-mail: hyderabad@newagepublishers.com
- **Kolkata** RDB Chambers (Formerly Lotus Cinema) 106A, 1st Floor, S N Banerjee Road, Kolkata-700 014
  Tel.: (033) 22273773, **Telefax:** 22275247, **E-mail: kolkata@newagepublishers.com**
- **Lucknow** 16-A, Jopling Road, Lucknow-226 001, **Tel.:** (0522) 2209578, 4045297, **Telefax:** 2204098
  E-mail: lucknow@newagepublishers.com
- **Mumbai** 142C, Victor House, Ground Floor, N.M. Joshi Marg, Lower Parel, Mumbai-400 013
  Tel.: (022) 24927869, **Telefax:** 24915415, **E-mail: mumbai@newagepublishers.com**
- **New Delhi** 22, Golden House, Daryaganj, New Delhi-110 002, **Tel.:** (011) 23262368, 23262370, **Telefax:** 43551305
  E-mail: sales@newagepublishers.com

ISBN: 978-81-224-3690-7

C-17-01-10000

Printed in India at Glorious Printers, Delhi.
Typeset at In-house, Delhi.

PUBLISHING GLOBALLY
NEW AGE INTERNATIONAL (P) LIMITED, PUBLISHERS
7/30 A, Daryaganj, New Delhi-110002
Visit us at **www.newagepublishers.com**
(CIN: U74899DL1966PTC004618)

PREFACE TO THE SECOND EDITION

The author has been encouraged with the enthusiastic response from the academic community for the first edition of his book "BASIC GEOTECHNICAL EARTHQUAKE ENGINEERING".

In consistent with the subject matter of the book and considering their relevance, it has been decided to add three new chapters for the second edition of the book. The format for these new chapters is same as that of first eleven chapters of the first edition of the book. Home work problems have not been at the end of Chapter 14, because it is a case study chapter.

I am sure that this second enlarged edition will also be received well by the academic fraternity.

KAMALESH KUMAR

PREFACE TO THE FIRST EDITION

Earthquake resistant geotechnical construction has become an important design aspect recently. This book BASIC GEOTECHNICAL EARTHQUAKE ENGINEERING is intended to be used as textbook for the beginners of the geotechnical earthquake engineering curriculum. Civil engineering undergraduate students as well as first year postgraduate students, who have taken basic undergraduate course on soil mechanics and foundation engineering, will find subject matter of the textbook familiar and interesting.

Emphasis has been given to the basics of geotechnical earthquake engineering as well as to the basics of earthquake resistant geotechnical construction in the textbook. At the end of each chapter home work problems have been given for practice. At appropriate places, solved numerical problems and exercise numerical problems have also been given to make the subject matter clear. Subject matter of the textbook can be covered in a course of one semester which is about of 4 to 4.5 months duration. List of references given at the end of book enlists references which have been used to prepare this basic book on geotechnical earthquake engineering. Although the book is on geotechnical earthquake engineering, the last chapter of book is on earthquake resistant design of buildings, considering its significance in the context of earthquake resistant construction.

The ultimate judges of the book will be students, who will use the book to understand the basic concepts of geotechnical earthquake engineering.

Suggestions to improve the usefulness of the book will be gratefully received.

KAMALESH KUMAR

Contents

INTRODUCTION TO GEOTECHNICAL EARTHQUAKE ENGINEERING

1.1 INTRODUCTION

The effect of earthquake on people and their environment as well as methods of reducing these effects is studied in earthquake engineering. It is a new discipline, with most of the developments in the past 30 to 40 years. Most earthquake engineers have structural or geotechnical engineering background. This book covers geotechnical aspects of earthquake engineering.

Geotechnical earthquake engineering is an area within geotechnical engineering. It deals with the design and construction of projects in order to resist the effect of earthquakes. Geotechnical earthquake engineering requires an understanding of geology, seismology and earthquake engineering. Furthermore, practice of geotechnical earthquake engineering also requires consideration of social, economic and political factors. In seismology, internal behavior of the earth as well as nature of seismic waves generated by earthquake is studied.

In geology, geologic data and principles are applied so that geologic factors affecting the planning, design, construction and maintenance of civil engineering works are properly recognized and utilized. Primary responsibility of geologist is to determine the location of fault, investigate the fault in terms of either active or passive, as well as evaluate historical records of earthquakes and their impact on site. These studies help to define design earthquake parameters. The important design earthquake parameters are peak ground accleration and magnitude of anticipated earthquake.

The very first step in geotechnical earthquake engineering is to determine the dynamic loading from the anticipated earthquake. The anticipated earthquake is also called design earthquake. For this purpose, following activities needs to be performed by geotechnical earthquake engineer:

- Investigation for the possibility of liquefaction at the site. Liquefaction causes complete loss of soil shear strength, causing bearing capacity failure, excessive settlement or slope movement. Consequently, this investigation is necessary.
- Calculation of settlement of structure caused by anticipated earthquake.
- Checking the bearing capacity and allowable soil bearing pressures, to make sure that foundation does not suffer a bearing capacity failure during the design earthquake.
- Investigation for slope stability due to additional forces imposed due to design earthquake. Lateral deformation of slope also needs to be studied due to anticipated earthquake.
- Effect of earthquake on the stability of retaining walls.
- Analyze other possible earthquake effects, such as surface faulting and resonance of the structure.
- Development of site improvement techniques to mitigate the effect of anticipated earthquake. These include Ground stabilization and ground water control.
- Determination of the type of foundation (shallow or deep), best suited for resisting the effect of design earthquake.
- To assist the structural engineer by investigating the effect of ground movement due to seismic forces on the structure.

1.2 EARTHQUAKE RECORDS

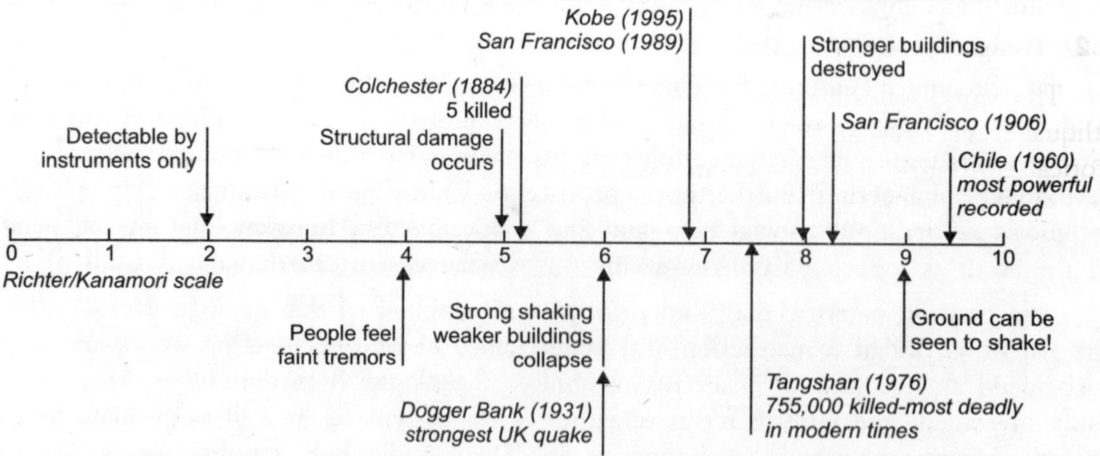

Fig. 1.1 Earthquake records (Courtesy: http://www.stvincet.ac.uk)

Accurate records of earthquake magnitudes have been kept only for some 100 years since the invention of the seismograph in the 1850s. Recent records of casualties are likely to be more reliable than those of earlier times. There are estimated to be some 500,000 seismic events each year. Out of these, about 100,000 can be felt and about 1,000 cause some form of damage. Some of the typical earthquake records have been shown in Fig. 1.1.

1.2.1 Most Powerful Earthquakes

Each increase of earthquake of 1 point on the Richter scale represents an increase of 10 times in the disturbance and a release of 30 times more energy. Richter scale is used to measure magnitude of earthquake and has been discussed in detail later in the book. The smallest measurable events associated with earthquake release energy in the order of 20J. This is equivalent to dropping a brick from a table top. The most powerful recorded earthquake was found to release energy which is equivalent to the simultaneous detonation of 50 of the most powerful nuclear bombs. Most powerful historical earthquakes are shown in Table 1.1.

Table 1.1 Most Powerful Historical Earthquakes (Courtesy: http://www.stvincet.ac.uk)

Year	*Location*	*Magnitude*	*Persons killed*
1960	Chile	9.3	22,000
1964	Alaska	9.2	130
1952	Kamchatka	9.0	0
1965	Aleutian Islands	8.7	0
1922	Chile	8.7	0
1957	Aleutian Islands	8.6	0
1950	Himalayan region	8.6	8,500
1906	Ecuador	8.6	500
1963	Kurile Islands	8.6	0
1923	Alaska	8.5	0

1.2.2 Deadliest Earthquakes

The world's deadliest earthquake may have been the great Honan Shensi province earthquake in China, in 1556. Estimates put the total death toll at 830,000. Most deadliest historical earthquakes are shown in Table 1.2.

Table 1.2 Most Deadliest Historical Earthquakes (Courtesy: http://www.stvincet.ac.uk)

Year	*Location*	*Magnitude*	*Persons killed*
1976	Tangshan, China	7.5	655,000
1927	Qinghai, China	7.7	200,000
1923	Tokyo, Japan	7.9	143,000
1908	Messina, Italy	6.9	110,000
1920	Northern China	8.3	100,000
1932	Gansu, China	7.6	70,000
1970	Peru	8.0	54,000
1990	Iran	7.9	50,000
1935	Quetta, Pakistan	8.1	30,000
1939	Erzincan, Turkey	7.7	30,000

Similar magnitude earthquakes may result in widely varying casualty rates. For example, the San Francisco Loma Prieta earthquake of 1989, left 69 people dead. On the other hand, the Azerbaijan earthquake, left some 20,000 killed. Both earthquakes measured 6.9 on the Richter scale. The differences are partly explained by the quality of building and civil disaster preparations of the inhabitants in the San Francisco area.

1.3 EARTHQUAKE RECORDS OF INDIA

Throughout the invasions of different ethnic and religious entitites in the past two millennia the Indian subcontinent has been known for its unique isolation imposed by surrounding mountains and oceans. The northern, eastern and western mountains are the boundaries of the Indian plate. The shorelines indicate ancient plate boundaries. Initially Indian subcontinent was a single Indian plate. Only in recent time have the separate nations of Pakistan, India, and Bangladesh have come up within Indian plate.

Surprisingly, despite a written tradition extending beyond 1500 BC, very little is known about Indian earthquakes earlier than 500 years before the present. Actually, records are close to complete only for earthquakes in the most recent 200 years. This presents a problem for estimating recurrence intervals between significant earthquakes. Certainly no repetition of an earthquake has ever been recognized in the written record of India. However, great earthquakes in the Himalaya are found to do so at least once and possibly as much as three times each millennium. The renewal time for earthquakes in the Indian sub-continent exceeds many thousands of years. Consequently, it is unlikely that earthquakes will be repeated during the time of written records.

However, trench investigations indicate that faults have been repeatedly active on the subcontinent (Sukhija et al., 1999; Rajendran, 2000) as well as within the Himalayan plate boundary (Wesnousky et al., 1999). The excavation of active faults and liquefaction features play important role in extending historic earthquake record of Indian earthquakes in the next several decades.

1.3.1 Tectonic Setting of India

India is currently penetrating into Asia at a rate of approximately 45 mm/year. Furthermore, it is also rotating slowly anticlockwise (Sella et al., 2002). This rotation and translation results in left-lateral transform slip in Baluchistan at approximately 42 mm/year as well as right-lateral slip relative to Asia in the Indo-Burman ranges at 55 mm/year (Fig. 1.2). Since, structural units at its northern, western and eastern boundaries are complex, these velocities are not directly observable across any single fault system. Deformation within Asia reduces India's convergence with Tibet to approximately 18 mm/year (Wang et al., 2001). However, since Tibet is extending east-west, convergence across the Himalaya is approximately normal to the arc. Arc-normal convergence across the Himalaya results in the development of potential slip available to drive large thrust earthquakes beneath the Himalaya at roughly 1.8 m/century. Consequently, earthquakes associated with, 6m of slip (say) cannot occur before the elapse of an interval of at least three centuries (Bilham et al., 1998).

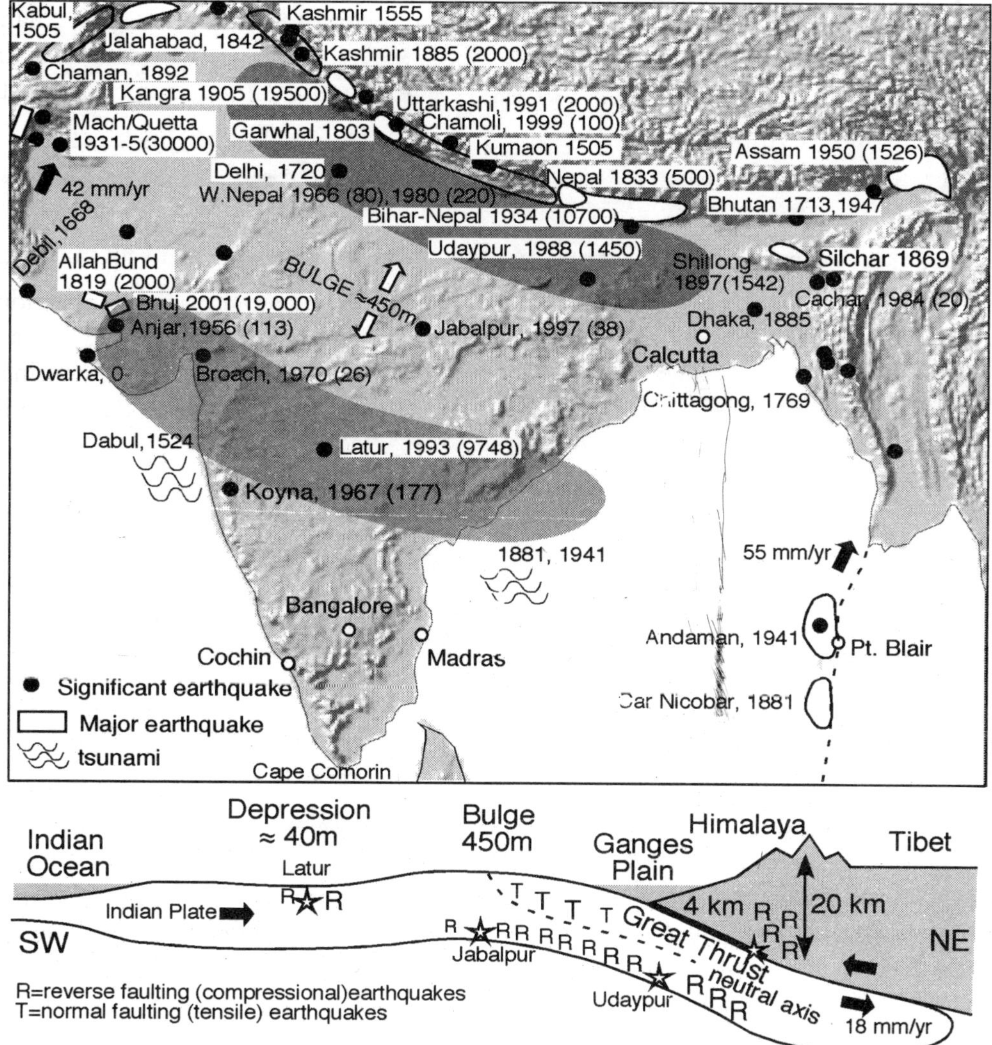

Fig. 1.2 Schematic views of Indian tectonics. Plate boundary velocities are indicated in mm/year. Shading indicates flexure of India: a 4 km deep trough near the Himalaya and an inferred minor (40 m) trough in south central India are separated by a bulge that rises approximately 450 m. Tibet is not a tectonic plate: it extends east-west and converges north-south at approximately 12 mm/year. At the crest of the flexural bulge the surface of the Indian plate is in tension and its base is in compression. Locations and dates of important earthquakes mentioned in the text are shown, with numbers of fatalities in parenthesis where known. With the exception of the Car Nicobar 1881, Assam 1897 and Bhuj 2001 events, none of the rupture zones major earthquakes are known with any certainty. The estimated rupture zones of pre-1800 great earthquakes are shown as unfilled outlines, whereas more recent events are filled white. (Courtesy: <http://cires.colorado.edu>)

GPS measurements in India reveal that convergence is less than 5±3 mm/year from Cape Comorin (Kanya Comori) to the plains south of the Himalaya (Paul et al., 2001). Consequently, Indian Plate is not expected to host frequent seismicity. However, collision of

India has resulted in flexure of Indian Plate (Bilham et al., 2003). The wavelength of flexure is of the order of 650 km. It results in approximately 450-m-high bulge near the central Indian Plateau. Normal faulting earthquakes occur north of this flexural bulge (e.g. possibly on 15 July 1720 near Delhi) as well as deep reverse faulting also occurs beneath its crest (e.g. the May 1997 Jabalpur earthquake). Furthermore, shallow reverse faulting also occurs south of the flexural bulge where the Indian plate is depressed (e.g. the Sept. 1993 Latur earthquake, Fig. 1.2).

The presence of flexural stresses as well as of plate-boundary slip permits all mechanisms of earthquakes to occur beneath the Lesser Himalaya (Fig. 1.2). At depths of 4 – 18 km great thrust earthquakes with shallow northerly dip occur infrequently. This permits the northward descent of the Indian Plate beneath the subcontinent. Earthquakes in the Indian Plate beneath these thrust events range from tensile just below the plate interface, to compressional and strike-slip at depths of 30-50 km (e.g. the August 1988 Udaypur earthquake).

A belt of microearthquakes and moderate earthquakes beneath the Greater Himalaya on the southern edge of Tibet indicates a transition from stick-slip fault to aseismic creep at around 18 km. This belt of microseismicity defines a small circle which has a radius of 1695 km (Seeber and Gornitz, 1983).

1.3.2 Historic Data Sources and Catalogues

Early earthquakes described in mythical terms include extracts in the Mahabharata during the Kurukshetra battle (Iyengar, 1994). There are several semi-religious texts mentioning a probable Himalayan earthquake during the time of enlightment of Buddha c. 538 BC.

Archaeological excavations in Sindh and Gujarat suggest earthquake damage to now abandoned Harrappan cities. A probable earthquake around 0 AD near the historically important city of Dwarka is recorded, since zones of liquefaction in the archeological excavations of the ancient city were found (Rajendran et al., 2003). The town of Debal (Dewal, Debil, Diul Sind or Sindi) near the current site of Karachi was alleged to have been destroyed in 893 AD (Oldham 1883). Rajendran and Rajendran (2002) present a case that the destruction of Debil was caused by an earthquake linked to the same fault system responsible for the 1819 and 2001 Rann of Kachchh earthquakes. However, Ambraseys (2003) notes that the sources of Oldham's account probably refer to Daibul (Dvin) in Armenia, and that liquefaction 1100 years ago must be attributed to a different earthquake.

There was a massive earthquake in the Kathmandu Valley in 1255 (Wright, 1877). It was a great earthquake because it was alleged to have been followed by three years of aftershocks. However, the absence of reports from other locations renders this of little value in estimating its rupture dimensions or magnitude. Similarly the arrival of Vasco de Gama's fleet in 1524 coincided with a violent sea-quake and tsunami that caused alarm at Dabul (Bendick and Bilham, 1999). Note that this Portuguese port on the Malabar Coast is unrelated to Debil above.

An important recent realization is that a sequence of significant earthquakes occurred throughout the west Himalaya in the 16th century. The sequence started in Kashmir in 1501, which was followed by two events a month apart in Afghanistan and in the central Himalaya.

The sequence concluding with a large earthquake in Kashmir in 1555. The central Himalayan earthquake may have been based on its probable rupture area. It destroyed monasteries along a 500 km segment of southern Tibet, in addition to demolishing structures in Agra and other towns in northern India.

A Himalayan earthquake that damaged the Kathmandu Valley in 1668 is mentioned briefly in Nepalese histories. Earthquakes in the 18th century are poorly documented. An earthquake near Delhi in 1720 caused damage and apparent liquefaction. However, little else is known of this event (Kahn 1874; Oldham 1883). This event, from its location, appears to be a normal faulting event. However, since there is absence of damage accounts from the Himalaya it may have been a Himalayan earthquake as well. In 1713 a severe earthquake damaged Bhutan and parts of Assam (Ambraseys and Jackson, 2003).

Thirteen years later, in September 1737, a catastrophic earthquake is alleged to have occurred in Calcutta. This is the most devastating earthquake to be listed in many catalogues of Indian as well as in global earthquakes. There was a storm surge that resulted in numerous deaths by drowning along the northern coast of the Bay of Bengal. The hand-written ledgers of the East India Company in Bengal detail storm and flood damage to shipping, warehouses and dwellings in Calcutta (Bilham, 1994).

India in the early 19th century was as yet incompletely dominated by a British colonial administration. An earthquake in India was something of a rarity. It generated detailed letters from residents describing its effects. Few of the original letters have survived, but the earthquakes in Kumaon in 1803, Nepal in 1833 and Afghanistan in 1842 were felt sufficiently widely to lead scientifically inclined officials to take a special interest in the physics and geography of earthquakes.

An army officer named Baird-Smith wrote a sequence of articles 1843-1844 in the Asiatic Society of Bengal summarizing data from several Indian earthquakes and venturing to offer explanations for their occurrence. He was writing shortly after the first Afghan war which had coincided with a major 1842 earthquake in the Kunar Valley of NE Afghanistan (Ambraseys and Bilham, 2003a). The director of the Geological Survey of India, Thomas Oldham (1816-1878) published the first real catalog of significant Indian events in 1883. His catalog includes earthquakes from 893 to 1869.

His son, Richard D. Oldham (1858-1936), wrote accounts of four major Indian earthquakes (1819, 1869, 1881, and 1897). He completed first his father's manuscript on the 1869 Silchar, Cachar, Assam earthquake which was published under his father's name. He next investigated the December 1881 earthquake in the Andaman Islands, visiting and mapping the geology of some of the islands. His account of the 1897 Shillong Plateau earthquake in Assam was exemplary, and according to Richter provided the best available scientific analyses of available physical data on any earthquake at that time.

R.D. Oldham's accounts established a template for the study of earthquakes that occurred in India subsequently. The great earthquakes of 1905 Kangra and 1934 Bihar/Nepal were each assigned to Geological Survey of India special volumes. However, these never quite matched the insightful observations of Oldham's 1899 volume. Investigations of the yet larger Assam earthquake of 1950 were published as a compilation undertaken by separate investigators (e.g. Ray 1952 and Tandon, 1952). In many ways this proved to be the least conclusive of the studies of the 5 largest Indian earthquakes during 1819-1950.

Home Work Problems

1. Explain the concept of geotechnical earthquake engineering.
2. Enlist activities to be performed by geotechnical earthquake engineer.
3. Write short note on tectonic setting of India.
4. Using historic data sources explain about historic earthquakes in India.

EARTHQUAKES

2.1 PLATE TECTONICS, THE CAUSE OF EARTHQUAKES

The plates consist of an outer layer of the Earth. This is called the **lithosphere**. It is cool enough to behave as a more or less rigid shell. Occasionally the hot **asthenosphere** of the Earth finds a weak place in the lithosphere to rise buoyantly as a plume, or hotspot. The satellite image in Fig. 2.1 below shows the volcanic islands of the Galapagos hotspot.

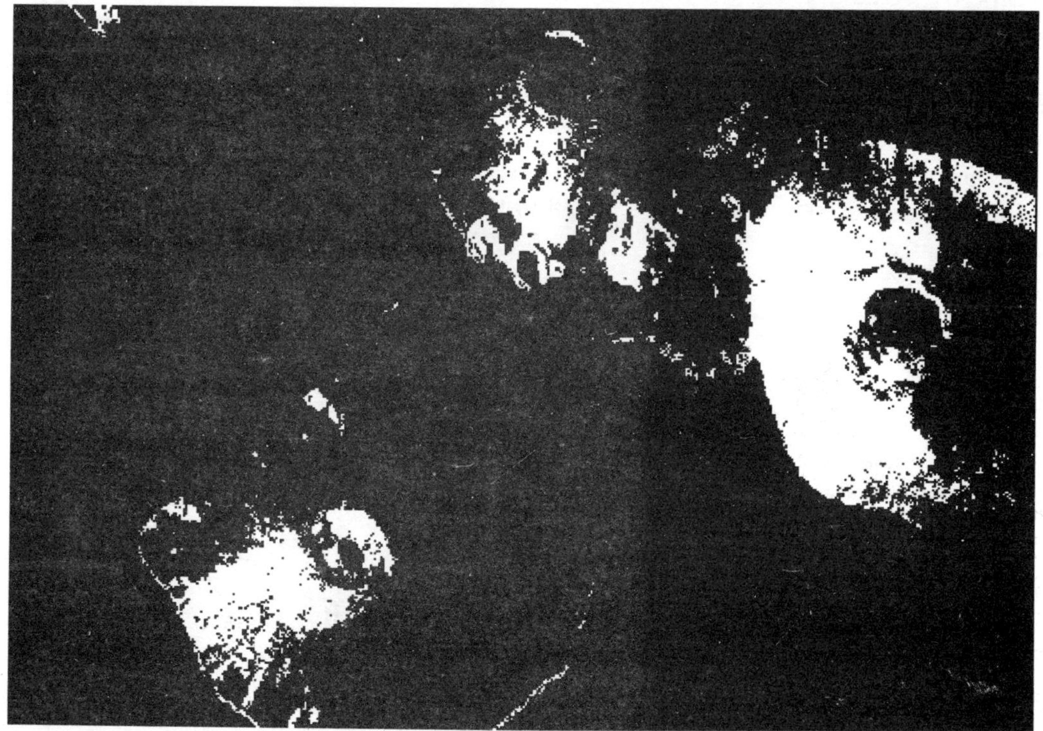

Fig. 2.1 Volcanic islands (Courtesy: NASA)

Only lithosphere has the strength and the brittle behavior to fracture in an earthquake. The map below (Fig. 2.2) locates earthquakes around the globe. They are not evenly distributed; the boundaries between the plates grind against each other, producing most earthquakes. Consequently, the help lines of earthquakes define the plates.

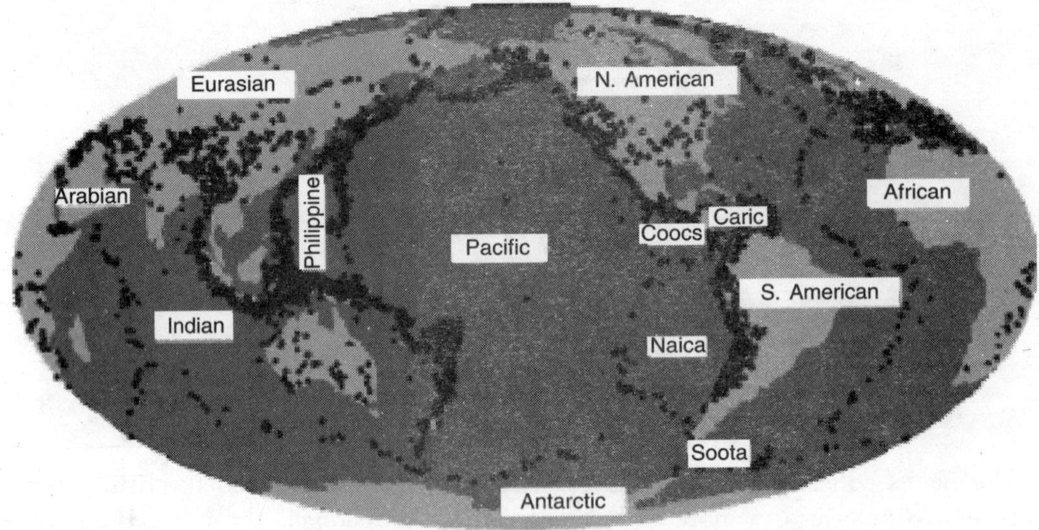

Fig. 2.2 Tectonic plates (Courtesy: USGS)

In cross section, the Earth releases its internal heat by convecting, or boiling much like a pot of pudding on the stove. Hot asthenospheric mantle rises to the surface and spreads laterally, transporting oceans and continents as on a slow conveyor belt. The speed of this motion is a few centimeters per year, about as fast as your fingernails grow. The new lithosphere, created at the ocean spreading centers, cools as it ages and eventually becomes dense enough to sink back into the mantle. The subducted crust releases water to form volcanic island chains above, and after a few hundred million years will be heated and recycled back to the spreading centers.

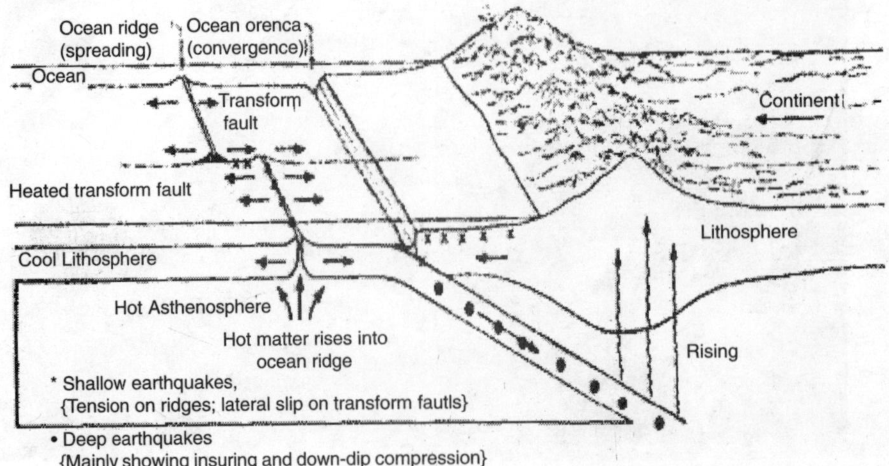

Fig. 2.3 Earthquake occurrence in different plate tectonic settings (Courtesy: http://seismo.unr.edu)

The map in (Fig. 2.3) of Earth's solid surface shows many of the features caused by plate tectonics. The oceanic ridges are the asthenospheric spreading centers, creating new oceanic crust. Subduction zones appear as deep oceanic trenches. Most of the continental mountain belts occur where plates are pressing against one another.

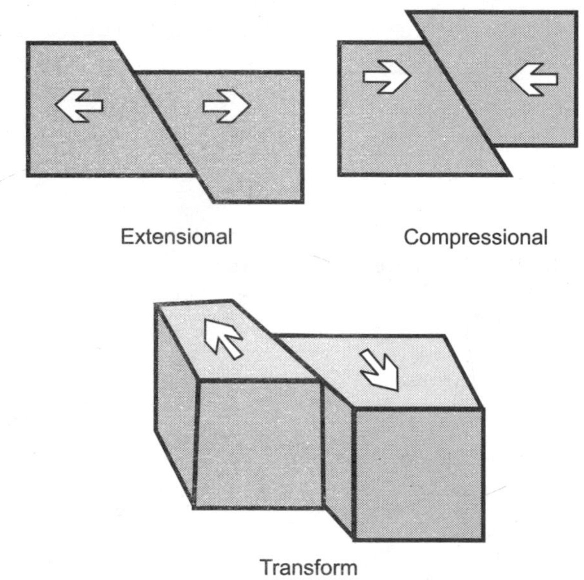

Fig. 2.4 Plate tectonic environments (Courtesy: http://seismo.unr.edu)

There are three main plate tectonic environments (Fig. 2.4): extensional, transform, and compressional. Plate boundaries in different localities are subject to different inter-plate stresses, producing these three types of earthquakes. Each type has its own special hazards.

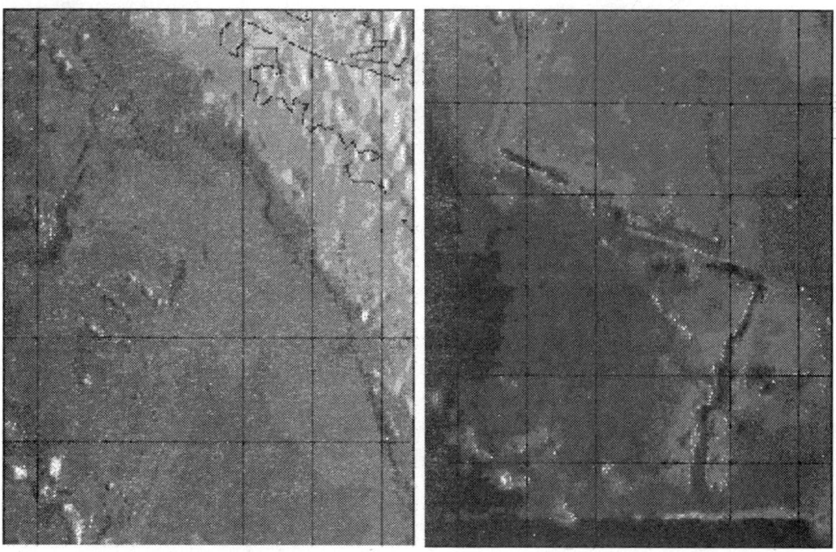

Fig. 2.5 Juan de Fuca spreading ridge (Courtesy: http://seismo.unr.edu)

At spreading ridges, or similar extensional boundaries, earthquakes are shallow. They are aligned strictly along the axis of spreading, and show an extensional mechanism. Earthquakes in extensional environments tend to be smaller than magnitude 8 (magnitude of earthquake has been discussed in detail later).

A close-up topographic picture (Fig. 2.5) of the Juan de Fuca spreading ridge, offshore of the Pacific Northwest, shows the turned-up edges of the spreading center. As crust moves away from the ridge it cools and sinks. The lateral offsets in the ridge are joined by the transform faults.

A satellite view (Fig. 2.6) of the Sinai shows two arms of the Red Sea spreading ridge, exposed on land.

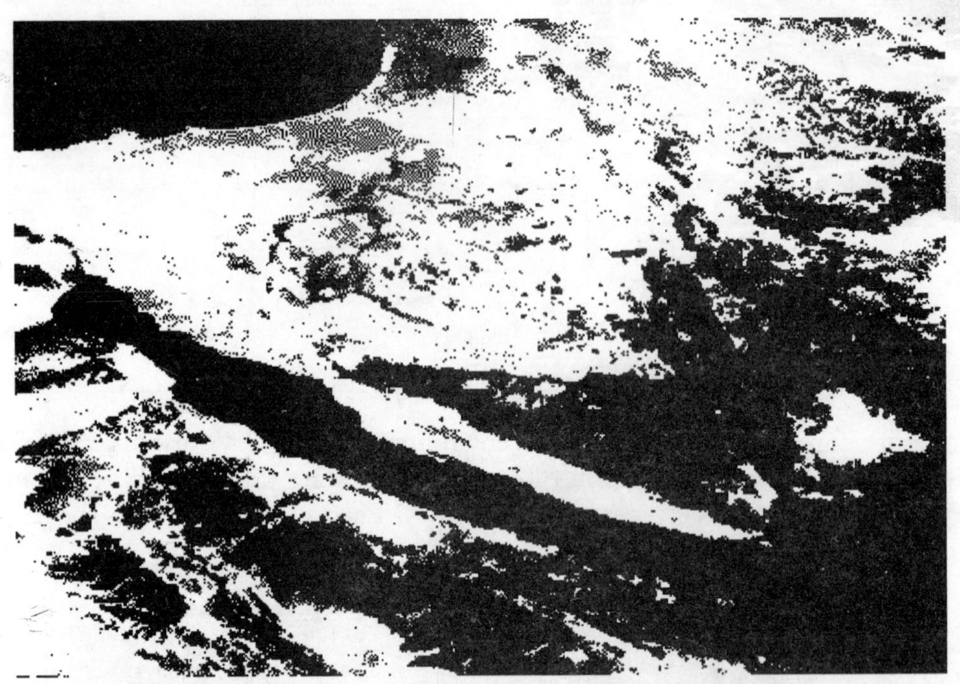

Fig. 2.6 Two arms of red sea spreading ridge (Courtesy: NASA)

Extensional ridges exist elsewhere in the solar system, although they never attain the globe-encircling extent the oceanic ridges have on Earth. This synthetic perspective of a large volcano on Venus (Fig. 2.7) is looking up the large rift on its flank.

At transforms, earthquakes are shallow, running as deep as 25 km. The mechanisms indicate strike-slip motion. Transforms tend to have earthquakes smaller than magnitude 8.5.

The San Andreas fault (Fig. 2.8) in California is a nearby example of a transform, separating the Pacific from the North American plate. At transforms the plates mostly slide past each other laterally, producing less sinking or lifing of the ground than extensional or compressional environments. The white dots in Fig. 2.8 locate earthquakes along strands of this fault system in the San Francisco Bay area.

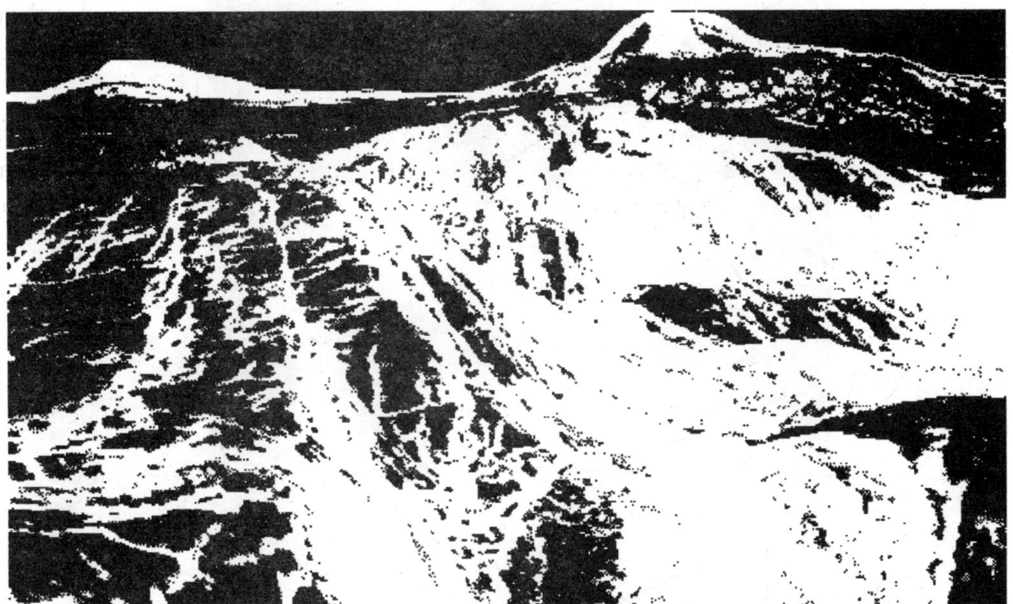

Fig. 2.7 Large volcano on Venus (Courtesy: NASA/JPL)

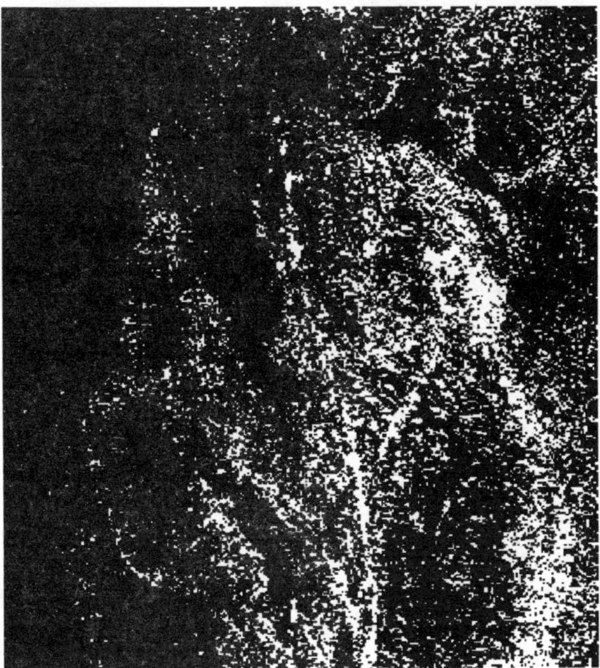

Fig. 2.8 The San Andreas fault in California (Courtesy: USGS)

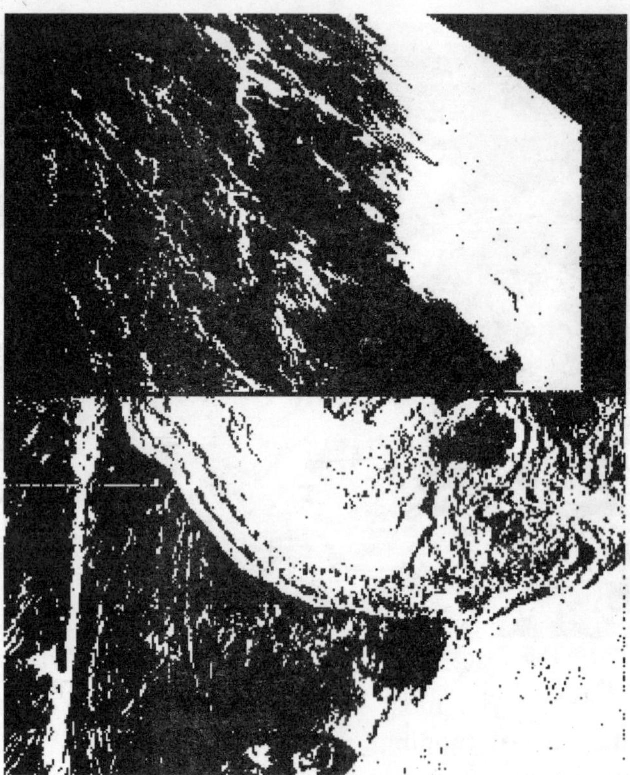

Fig. 2.9 (Courtesy: NASA, Topography from NOAA)

At compressional boundaries, earthquakes are found in several settings ranging from the very near surface to several hundred kilometers depth, since the coldness of the subducting plate permits brittle failure down to as much as 700 km. Compressional boundaries host Earth's largest quakes, with some events on subduction zones in Alaska and Chile having exceeded magnitude 9.

This oblique orbital view of Fig. 2.9 looking east over Indonesia shows the clouded tops of the chain of large volcanoes. The topography of Fig. 2.9 shows the Indian plate, streaked by hotspot traces and healed transforms, subducting at the Javan Trench.

Sometimes continental sections of plates collide; both are too light for subduction to occur. The satellite image (Fig. 2.10) below shows the bent and rippled rock layers of the Zagros Mountains in southern Iran, where the Arabian plate is impacting the Iranian plate.

Nevada has a complex plate-tectonic environment, dominated by a combination of extensional and transform motions. The Great Basin shares some features with the great Tibetan and Anatolian plateaus. All three have large areas of high elevation, and show varying amounts of rifting and extension distributed across the regions. This is unlike oceanic spreading centers, where rifting is concentrated narrowly along the plate boundary. The numerous north-south mountain ranges that dominate the landscape from Reno to Salt Lake City are the consequence of substantial east-west extension, in which the total extension may be as much as a factor of two over the past 20 million years.

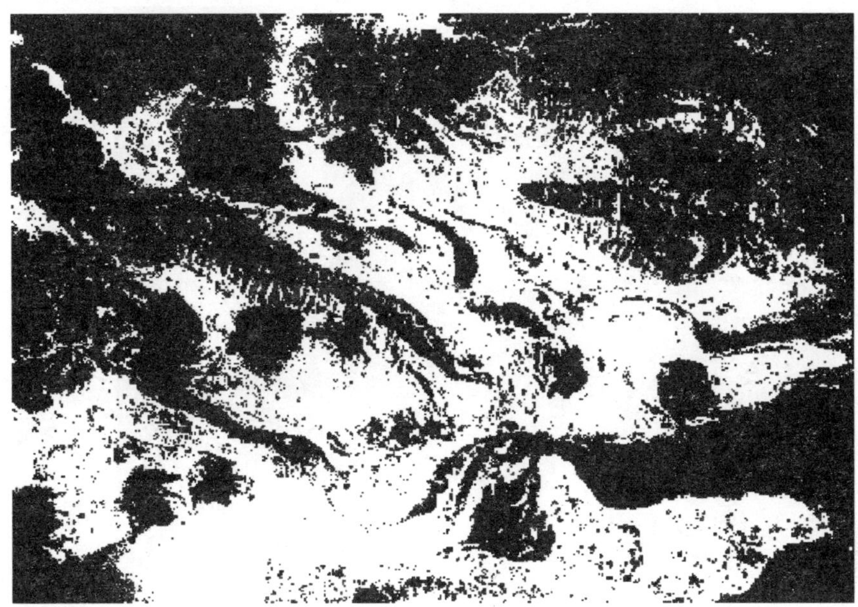

Fig. 2.10 The Zagros Mountains in southern Iran (Courtesy: NASA)

The extension seems to be most active at the eastern and western margins of the region, i.e. the mountain fronts running near Salt Lake City and Reno. The western Great Basin also has a significant component of shearing motion superimposed on this rifting. This is part of the Pacific - North America plate motion. The total motion is about 5 cm/year. Of this, about 4 cm/year takes place on the San Andreas fault system near the California coast, and the remainder, about 1 cm/year, occurs east of the Sierra Nevada mountains, in a zone geologists know as the Walker Lane.

As a result, Nevada hosts hundreds of active extensional faults, and several significant transform fault zones as well. While not as actively or rapidly deforming as the plate boundary in California, Nevada has earthquakes over much larger areas. While some regions in California, such as the western Sierra Nevada, appear to be isolated from earthquake activity, earthquakes have occurred everywhere in Nevada.

2.2 SEISMIC WAVES

When an earthquake occurs, different types of seismic waves are produced. The main seismic wave types are Compressional (P), Shear (S), Rayleigh (R) and Love (L) waves. P and S waves are often called body waves because they propagate outward in all directions from a source (such as an earthquake) and travel through the interior of the Earth. Love and Rayleigh waves are surface waves and propagate approximately parallel to the Earth's surface. Although surface wave motion penetrates to significant depth in the Earth, these types of waves do not propagate directly through the Earth's interior. Descriptions of wave characteristics and particle motions for the four wave types are given in Table 2.1.

Table 2.1: Seismic Waves (Courtesy: http://web.ics.purdue.edu)

Wave Type (and names)	*Particle Motion*	*Typical Velocity*	*Other Characteristics*
P, Compressional, Primary, Longitudinal	Alternating compressions ("pushes") and dilations ("pulls") which are directed in the same direction as the wave is propagating (along the ray path); and therefore, perpendicular to the wavefront.	V_P ~ 5-7 km/s in typical Earth's crust; >~ 8 km/s in Earth's mantle and core; ~1.5 km/s in water; ~0.3 km/s in air.	P motion travels fastest in materials, so the P-wave is the first-arriving energy on a seismogram. Generally smaller and higher frequency than the S and Surface-waves. P waves in a liquid or gas are pressure waves, including sound waves.
S, Shear, Secondary, Transverse	Alternating transverse motions (perpendicular to the direction of propagation, and the ray path); commonly approximately polarized such that particle motion is in vertical or horizontal planes.	V_S ~ 3-4 km/s in typical Earth's crust; >~ 4.5 km/s in Earth's mantle; ~ 2.5-3.0 km/s in (solid) inner core.	S-waves do not travel through fluids, so do not exist in Earth's outer core (inferred to be primarily liquid iron) or in air or water or molten rock (magma). S waves travel slower than P waves in a solid and, therefore, arrive after the P wave.
L, Love, Surface waves, Long waves	Transverse horizontal motion, perpendicular to the direction of propagation and generally parallel to the Earth's surface.	V_L ~ 2.0-4.4 km/s in the Earth depending on frequency of the propagating wave, and therefore the depth of penetration of the waves. In general, the Love waves travel slightly faster than the Rayleigh waves.	Love waves exist because of the Earth's surface. They are largest at the surface and decrease in amplitude with depth. Love waves are dispersive, that is the wave velocity is dependent on frequency, generally with low frequencies propagating at higher velocity. Depth of penetration of the Love waves is also dependent on frequency, with lower frequencies penetrating to greater depth.
R, Rayleigh, Surface waves, Long waves, Ground roll	Motion is both in the direction of propagation and perpendicular (in a vertical plane), and "phased" so	V_R ~ 2.0-4.2 km/s in the Earth depending on frequency of the propagating	Rayleigh waves are also dispersive and the amplitudes generally decrease with depth in the Earth. Appearance and particle motion

	that the motion is generally elliptical-either prograde or retrograde.	wave, and therefore the depth of penetration of the waves.	are similar to water waves. Depth of penetration of the Rayleigh waves is also dependent on frequency, with lower frequencies penetrating to greater depth.

2.3 FAULTS

The outer part of the Earth is relatively cold. So when it is stressed it tends to break, particularly if pushed quickly! These breaks, across which slip has occurred, are called faults. The most obvious manifestations of active faulting are earthquakes. Since these tend to happen along the boundaries between plates, this is where most of the active faulting occurs today. However, faulting can occur in the middle of the plates too, particularly in the continents. In general, faulting is restricted to the top 10-15 km of the Earth's crust. Below this level other things happen.

There is a wide range of faulting. Furthermore, faults themselves can form surprisingly complex patterns. Different types of faults tend to form in different settings. It has been found that the faults at active rifts are different from those along the edges of mountain ranges. Consequently, understanding the types and patterns of ancient fault can help geologists to predict and reconstruct the forms of ancient rifts and mountain ranges. The faulting patterns can have enormous economic importance. Faults can control the movement of groundwater. They can exert a strong influence on the distribution of mineralisation and the subsurface accumulations of hydrocarbons. Furthermore, they can have a major influence on the shaping of the landscape. When an earthquake occurs only a part of a fault is involved in the rupture. That area is usually outlined by the distribution of aftershocks in the sequence.

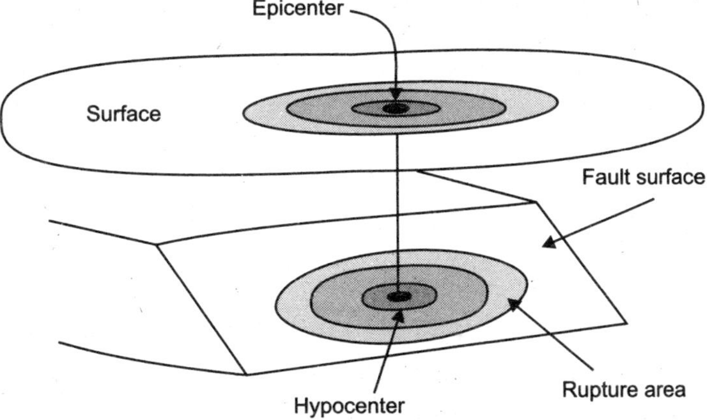

Fig. 2.11 Hypocenter and epicenter of earthquake (Courtesy: http://eqseis.geosc.psu.edu)

We call the "point" (or region) where an earthquake rupture initiates the **hypocenter** or **focus**. The point on Earth's surface directly above the hypocenter is called the **epicenter** (refer Fig. 2.11). When we plot earthquake locations on a map, we usually center the symbol representing an event at the epicenter.

Generally, the area of the fault that ruptures increases with magnitude. Some estimates of rupture area are presented in the Table 2.2 below.

Table 2.2: Rupture area of certain earthquakes (Courtesy: http:// eqseis.geosc.psu.edu)

Date	*Location*	*Length (km)*	*Depth (km)*	*(Mw)*
12/16/54	Dixie Peak, NV	42	14	6.94
06/28/66	Parkfield, CA	35	10	6.25
02/09/71	San Fernando Valley, CA	17	14	6.64
10/28/83	Borah Peak, ID	33	20	6.93
10/18/89	Loma Prieta, CA	40	16	6.92
06/28/92	Landers, CA	62	12	7.34

Although the exact area associated with a given size earthquake varies from place to place and event to event, we can make predictions for "typical" earthquakes based on the available observations (Refer Table 2.3 below). These numbers give a rough idea of the size of structure that we are talking about when we discuss earthquakes.

Table 2.3: Fault dimensions and earthquakes (Courtesy: http://eqseis.geosc.psu.edu)

Magnitude	*Fault Dimensions (Length × Depth, in km)*
4.0	1.2 × 1.2
5.0	3.3 × 3.3
6.0	10 × 10
6.5	16 × 16, 25 × 10
7.0	40 × 20, 50 × 15
7.5	140 × 15, 100 × 20, 72 × 30, 50 × 40, 45 × 45
8.0	300 × 20, 200 × 30, 150 × 40, 125 × 50

2.3.1 Fault Structure

Although the number of observations of deep fault structure is small, the available exposed faults provide some information on the deep structure of a fault. A fault "zone" consists of several smaller regions defined by the style and amount of deformation within them.

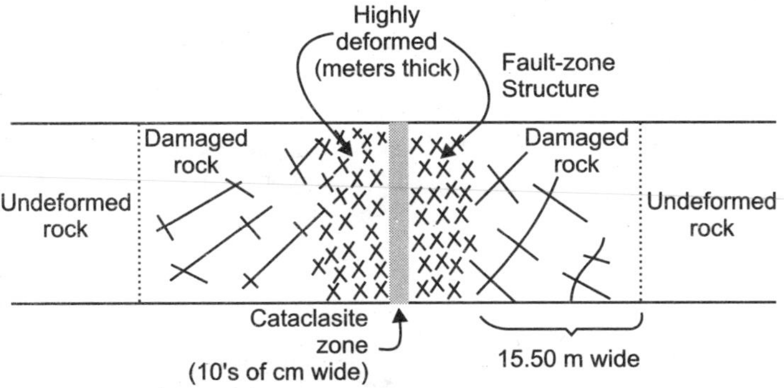

Fig. 2.12 Structure of an exposed section of a vertical strike-slip fault zone (after Chester et al., Journal of Geophysical Research, 1993).

Fig. 2.12 shows structure of an exposed section of a vertical strike-slip fault zone. The center of the fault is the most deformed and is where most of the offset or slip between the surrounding rock occurs. The region can be quite small, about as wide as a pencil is long, and it is identified by the finely ground rocks called cataclasite (we call the ground up material found closer to the surface, gouge). From all the slipping and grinding, the gouge is composed of very fine-grained material that resembles clay. Surrounding the central zone is a region several meters across that contains abundant fractures. Outside that region is another that contains distinguishable fractures, but much less dense than the preceding region. Last is the competent "host" rock that marks the end of the fault zone.

2.3.2 Fault Classifications

Active, Inactive, and Reactivated Faults

Active faults are structure along which we expect displacement to occur. By definition, since a shallow earthquake is a process that produces displacement across a fault. All shallow earthquakes occur on active faults.

Inactive faults are structures that we can identify, but which do not have earthquakes. As we can imagine, because of the complexity of earthquake activity, judging a fault to be inactive can be tricky. However, often we can measure the last time substantial offset occurred across a fault. If a fault has been inactive for millions of years, it's certainly safe to call it inactive. However, some faults only have large earthquakes once in thousands of years, and we need to evaluate carefully their hazard potential.

Reactivated faults form when movement along formerly inactive faults can help to alleviate strain within the crust or upper mantle. Deformation in the New Madrid seismic zone in the central United States is a good example of fault reactivation.

Faulting Geometry

Faulting is a complex process and the variety of faults that exists is large. We will consider a simplified but general fault classification based on the geometry of faulting, which we describe by specifying three angular measurements: dip, strike, and slip.

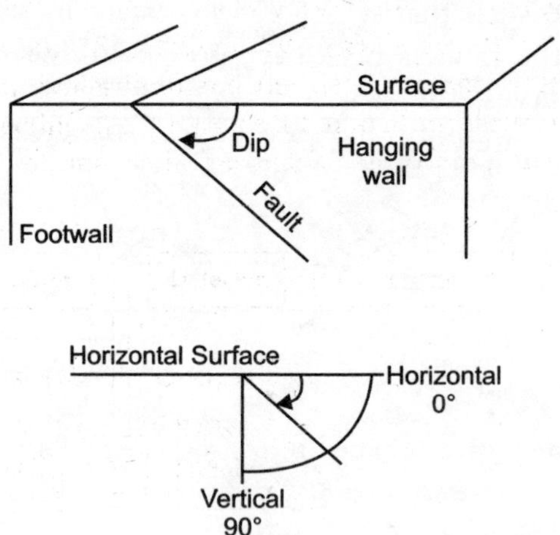

Fig. 2.13 Figure explaining about **dip** (Courtesy: http://eqseis.geosc.psu.edu)

In Earth, faults take on a range of orientations from vertical to horizontal. Dip is the angle that describes the steepness of the fault surface. This angle is measured from Earth's surface, or a plane parallel to Earth's surface. The dip of a horizontal fault is zero (usually specified in degrees: 0°), and the dip of a vertical fault is 90°. We use some old mining terms to label the rock "blocks" above and below a fault. If you were tunneling through a fault, the material beneath the fault would be by your feet, the other material would be hanging above your head. The material resting on the fault is called the hanging wall, the material beneath the fault is called the footwall.

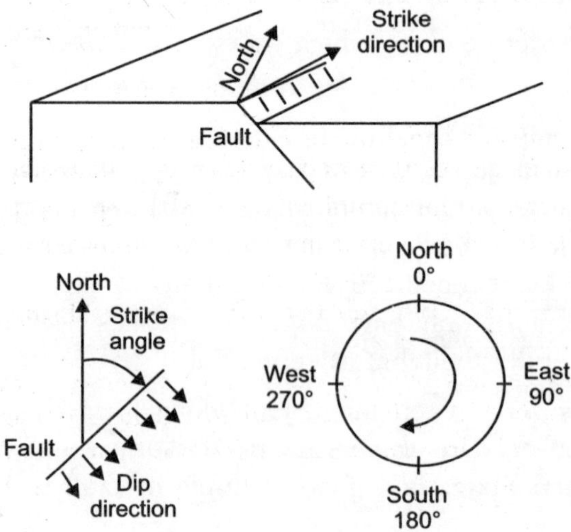

Fig. 2.14 Figure explaining about **strike** (Courtesy: http://eqseis.geosc.psu.edu)

The strike is an angle used to specify the orientation of the fault and measured clockwise from north. For example, a strike of 0° or 180° indicates a fault that is oriented in a north-south direction, 90° or 270° indicates east-west oriented structure. To remove the ambiguity, we always specify the strike such that when we "look" in the strike direction, the fault dips to our right. Of course if the fault is perfectly vertical we have to describe the situation as a special case. If a fault curves, the strike varies along the fault, but this seldom causes a communication problem if we are careful to specify the location (such as latitude and longitude) of the measurement.

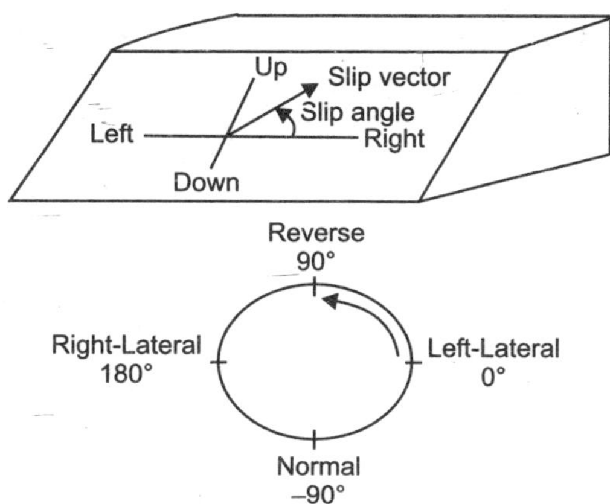

Fig. 2.15 Figure explaining about **slip** (Courtesy: http://eqseis.geosc.psu.edu)

Dip and strike describe the orientation of the fault, we also have to describe the direction of motion across the fault. That is, which way did one side of the fault move with respect to the other. The parameter that describes this motion is called the slip. The slip has two components, a "magnitude" which tells us how far the rocks moved, and a direction (it's a vector). We usually specify the magnitude and direction separately. The magnitude of slip is simply how far the two sides of the fault moved relative to one another. It is a distance usually a few centimeters for small earthquakes and meters for large events. The direction of slip is measured on the fault surface, and like the strike and dip, it is specified as an angle. Specifically the slip direction is the direction that the hanging wall moved relative to the footwall. If the hanging wall moves to the right, the slip direction is 0°; if it moves up, the slip angle is 90°, if it moves to the left, the slip angle is 180°, and if it moves down, the slip angle is 270° or –90°.

Hanging wall movement determines the geometric classification of faulting. We distinguish between "dip-slip" and "strike-slip" hanging-wall movements.

Dip-slip movement occurs when the hanging wall moved predominantly up or down relative to the footwall. If the motion was down, the fault is called a normal fault, if the movement was up, the fault is called a reverse fault. Downward movement is "normal" because we normally would expect the hanging wall to slide downward along the foot wall because of the pull of gravity. Moving the hanging wall up an inclined fault requires work to overcome friction on the fault and the downward pull of gravity.

When the hanging wall moves horizontally, it's a **strike-slip** earthquake. If the hanging wall moves to the left, the earthquake is called right-lateral, if it moves to the right, it's called a left-lateral fault. The way to keep these terms straight is to imagine that we are standing on one side of the fault and an earthquake occurs. If objects on the other side of the fault move to our left, it's a left-lateral fault, if they move to our right, it's a right-lateral fault.

When the hanging wall motion is neither dominantly vertical nor horizontal, the motion is called **oblique-slip**. Although oblique faulting isn't unusual, it is less common than the normal, reverse, and strike-slip movement. Fig. 2.16 explains about different fault classifications.

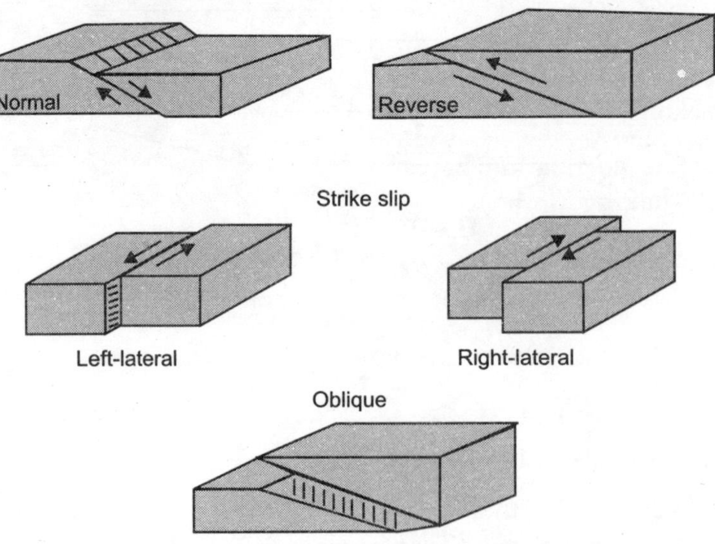

Fig. 2.16 Different fault classifications (Courtesy: http://eqseis.geosc.psu.edu)

Normal faulting is indicative of a region that is stretching, and on the continents, normal faulting usually occurs in regions with relatively high elevation such as plateaus. Reverse faulting reflects compressive forces squeezing a region and they are common in uplifting mountain ranges and along the coast of many regions bordering the Pacific Ocean. The largest earthquakes are generally low-angle (shallow dipping) reverse faults associated with "subduction" plate boundaries. Strike-slip faulting indicates neither extension nor compression, but identifies regions where rocks are sliding past each other. The San Andreas fault system is a famous example of strike-slip deformation-part of coastal California is sliding to the northwest relative to the rest of North America-Los Angeles is slowly moving towards San Francisco.

2.4 EARTHQUAKE MAGNITUDE AND INTENSITY

Magnitude of earthquake measures amount of energy released from the earthquake. Intensity of earthquake is based on damage to building as well as reactions of people. There are three commonly used magnitude scales to measure magnitude of earthquake. These have been explained below.

2.4.1 Local Magnitude Scale (M_L)

This scale is also called Richter scale. This scale is calculated as follows:

$$M_L = \log A - \log A_0 = \log A/A_0 \quad \text{...(2.1)}$$

where,

M_L = local magnitude (Richter magnitude scale)

A = maximum trace amplitude (in mm), as recorded by standard Wood-Anderson seismograph. The seismograph has natural period of 0.8 sec, damping factor of 80% and static magnification of 2800. It is located exactly 100 km from the epicenter.

A_0 = 0.001 mm. This corresponds to smallest earthquake that can be recorded.

Table 2.4 shows approximate correlation between local magnitude, peak ground acceleration and duration of shaking.

Table 2.4 Approximate correlation between local magnitude, peak ground acceleration and duration of shaking (g = acceleration due to gravity) (Courtesy: Day, 2002)

Local Magnitude (M_L)	*Typical peak ground acceleration a_{max} near the vicinity of the fault rupture*	*Typical duration of ground shaking near the vicinity of the fault rupture*
≤ 2	–	–
3	–	–
4	–	–
5	0.09g	2 sec
6	0.22g	12 sec
7	0.37g	24 sec
≥ 8	$\geq$ 0.50g	$\geq$ 34 sec

2.4.2 Surface Wave Magnitude Scale (M_s)

This scale is calculated as follows:

$$M_s = \log A' + 1.66 \log \Delta + 2.0 \quad ...(2.2)$$

where,

M_s = Surface wave magnitude scale.

A' = maximum ground displacement, μm.

Δ = epicenter distance to seismograph measured in degrees.

This magnitude scale is typically used for moderate to large earthquakes (having shallow focal depth). Furthermore, seismograph should be at least 1000 km from epicenter.

2.4.3 Moment Magnitude Scale (M_w)

In this scale, seismic moment M_0 is calculated first as follows:

$$M_0 = \mu A_f D \quad ...(2.3)$$

M_0 = seismic moment (N.m)

μ = shear modulus of material along fault plane (N/m^2). It has a value of 3×10^{10} N/m^2 for surface crust and 7×10^{12} N/m^2 for mantle.

A_f = area of fault plane undergoing slip, measured in m^2. (length of surface rupture times depth of aftershakes).

D = average displacement of ruptured segment of fault, measured in meters.

Moment magnitude scale M_w is interrelated with M_0 as follows:

$$M_w = -6.0 + 0.67 \log M_0 \quad ...(2.4)$$

This scale is found to work best for strike-slip faults.

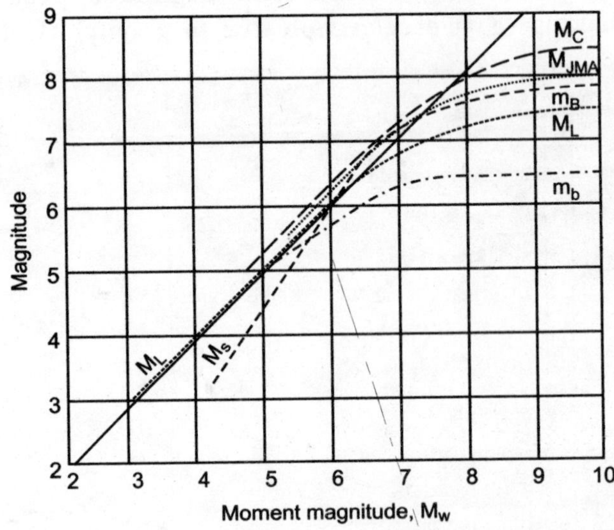

Fig. 2.17 Approximate relationships between the moment magnitude scale M_w and other magnitude scales (Courtesy: Day, 2002)

Approximate relation between different earthquake magnitude scales has been shown in Fig. 2.17. Based on the Fig. 2.17 it can be concluded that the magnitude scales M_L, M_s and M_w are reasonably close to each other below a value of about 7. At higher magnitude values, M_w tends to deviate from other two magnitude scales. Consequently, any of these three scales can be used to describe earthquake's magnitude for a magnitude value below about 7. For higher magnitudes, M_w is most suitable scale to describe earthquake's magnitude. Scales m_b, m_B and M_{JMA} given in Fig. 2.17 have not been discussed. All the magnitude scales tend to flatten out or get saturated at higher moment magnitude values. This saturation appears to occur when the ruptured fault dimension becomes much larger than the wavelength of seismic wave used in measuring the magnitude. M_L seems to become saturated at a value of about 7.3.

The intensity of an earthquake is based on observations of damaged structures. The intensity is also based on secondary effects like earthquake induced landslides, liquefaction, ground shaking, individual response etc. Intensity of earthquake can easily be determined in urban area. However, it is difficult to determine in rural area. Most commonly used intensity measurement scale is modified Mercalli intensity scale. This scale ranges from I to XII. I corresponds to a earthquake not felt. XII corresponds to a earthquake resulting in total destruction. Map containing contours of equal intensity is called isoseisms. In general the intensity will be highest in the general vicinity of the epicenter or at the location of maximum fault rupture. However, there can be local effects. The intensity is progressively found to decrease as the distance from the epicenter or from maximum fault rupture increases. The intensity scale can also be used to illustrate the anticipated damage at a site due to a future earthquake. Table 2.5 summarises the modified Mercalli intensity scale.

Table 2.5: Modified Mercalli Intensity Scale

Intensity Level	*Reaction of observers and types of damage*
I	**Reactions:** Not felt except by a few people under especially favorable circumstances. Damage: No damage.
II	**Reactions:** Felt only by a few persons at rest, especially on upper floors of buildings. Many people do not recogonize it as an earthquake. Damage: No damage. Delicately suspended objects may swing.
III	**Reactions:** Felt quite noticeably indoors, especially on upper floors of buildings. The vibration is like passing of a truck, and duration of the earthquake may be estimated. However, many people do not recogonize it as an earthquake. **Damage:** No damage. Standing motor cars may rock slightly.
IV	**Reactions:** During the day, felt indoors by many, outdoors by a few. At night, some people are awakened. The sensation is like a heavy truck striking the building. **Damage:** Dishes, windows and doors are disturbed. Walls make a cracking sound. Standing motor cars rock noticeably.
V	**Reactions:** Felt by nearly everyone, many awakened. Damage: Some dishes, windows, etc. broken. A few instances of cracked plaster and unstable objects overturned. Disturbances of trees, poles and other tall objects sometimes noticed. Pendulam clocks may stop.
VI	**Reactions:** Felt by everyone. Many people are frightened and run outdoors. Damage: There is slightly structural damage. Some heavy furntiture is moved, and there are some instances of fallen plaster or damaged chimneys.
VII	**Reactions:** Everyone runs outdoors. Noticed by persons driving motor cars. Damage: Negligible damage in buildings of good design and construction, slight to moderate damage in well-built ordinary structures, and considerable damage in poorly built or badly designed structures. Some chimneys are broken.

VIII	**Reactions:** Persons driving motor cars are disturbed. **Damage:** Slight damage in specially designed structures. Considerable damage in ordinary substantial buildings, with partial collapse. Great damage in poorly built structures. Panel walls are thrown out of frame structures. There is fall of chimneys, factory stacks, columns, monuments and walls. Heavy furniture is overturned. Sand and mud are ejected in small amounts, and there are changes in well-water levels.
IX	**Damage:** Considerable damage in specially designed structures. Well-designed frame structures are thrown out of plumb. There is great damage in substantial building with partial collapse. Buildings are shifted off of their foundations. The ground is conspicuously cracked, and underground pipes are broken.
X	**Damage:** Some well-built wooden structures are destroyed. Most masonry and framed structures are destroyed, including the foundations. The ground is badly cracked. There are bent train rails, a considerable number of landslides at river banks and steep slopes, shifted sand and mud, and water is splashed over their banks.
XI	**Damage:** Few, if any, masonry structures remain standing. Bridges are destroyed and train rails greatly bent. There are broad fissures in the ground, earth slumps and land slips in soft ground. Underground pipelines are out of service.
XII	**Reactions:** Waves seen on ground surface. Line of sight and level are distorted. Damage: Great to total damage of all construction work. Objects thrown upward into air.

2.5 SEISMOGRAPH

Seismologists study earthquakes by going out and looking at the damage caused by the earthquakes and by using seismographs. A **seismograph** is an instrument that records the shaking of the earth's surface caused by seismic waves.

The first seismograph was invented by the Chinese astronomer and mathematician Chang Heng. He called it an earthquake weathercock. Each of the eight dragons had a bronze ball in its mouth. Whenever there was even a slight earth tremor, a mechanism inside the seismograph would open the mouth of one dragon. The bronze ball would fall into the open mouth of one of the toads, making enough noise to alert someone that an earthquake had just happened. Imperial watchman could tell which direction the earthquake came from by seeing which dragon's mouth was empty.

Most seismographs today are electronic, but a basic seismograph is made of a drum with paper on it, a bar or spring with a hinge at one or both ends, a weight, and a pen. The one end of the bar or spring is bolted to a pole or metal box that is bolted to the ground. The weight is put on the other end of the bar and the pen is stuck to the weight. The drum with paper on it presses against the pen and turns constantly. When there is an earthquake, everything in the seismograph moves except the weight with the pen on it. As the drum and paper shake next to the pen, the pen makes squiggly lines on the paper, creating a record of the earthquake. This record made by the seismograph is called a **seismogram**. By studying the seismogram, the seismologist can tell how far away the earthquake was and how strong it was. This record doesn't tell the seismologist exactly where the epicenter was, just that the earthquake happened so many miles or kilometers away from that seismograph. To find the exact epicenter, we need to know what at least two other seismographs in other parts of the country or world recorded.

When we look at a seismogram, there will be wiggly lines all across it. These are all the seismic waves that the seismograph has recorded. Most of these waves were so small that nobody felt them. These tiny **microseisms** can be caused by heavy traffic near the seismograph, waves hitting a beach, the wind, and any number of other ordinary things that cause some shaking of the seismograph. There may also be some little dots or marks evenly spaced along the paper. These are marks for every minute that the drum of the seismograph has been turning. How far apart these minute marks are will depend on what kind of seismograph we have.

So which wiggles are the earthquake? The P wave will be the first wiggle that is bigger than the rest of the little ones (the microseisms). Because P waves are the fastest seismic waves, they will usually be the first ones that our seismograph records. The next set of seismic waves on your seismogram will be the S waves. These are usually bigger than the P waves. If there aren't any S waves marked on your seismogram, it probably means the earthquake happened on the other side of the planet. S waves can't travel through the liquid layers of the earth so these waves never made it to our seismograph.

The surface waves (Love and Rayleigh waves) are the other, often larger, waves marked on the seismogram. Surface waves travel a little slower than S waves (which are slower than P waves) so they tend to arrive at the seismograph just after the S waves. For shallow earthquakes (earthquakes with a focus near the surface of the earth), the surface waves may be the largest waves recorded by the seismograph. Often they are the only waves recorded a long distance from medium-sized earthquakes. Fig. 2.18 shows a typical seismogram.

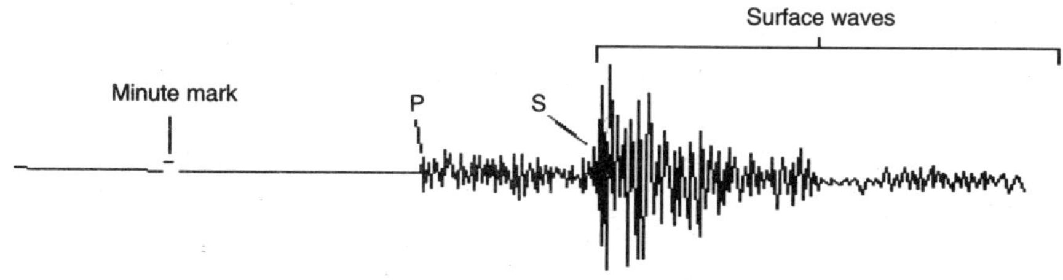

Fig. 2.18 Typical seismogram (Courtesy: http://www.geo.mtu.edu)

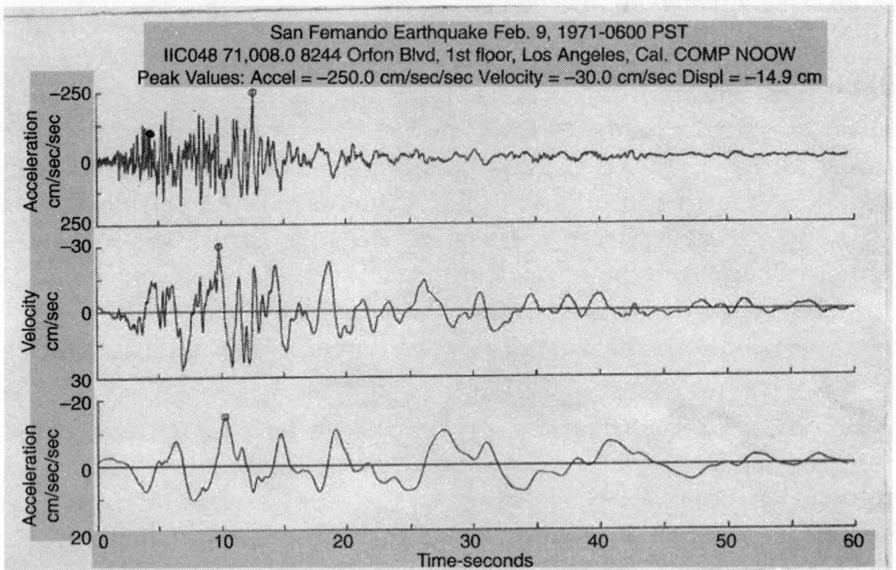

Fig. 2.19 Acceleration, velocity and displacement versus time recorded during San Fernando earthquake (Courtesy: Day, 2002)

The geotechnical earthquake engineer is often most interested in the peak ground acceleration a_{max} during the earthquake. An accelerograph is defined as a low magnification seismograph that is specially designed to record the ground acceleration during earthquake. Acceleration versus time plot obtained from accelerograph during San Fernando earthquake has been shown in Fig. 2.19. Fig. 2.19 also shows velocity versus time plot (obtained from integration of acceleration versus time plot), as well as displacement versus time plot (obtained from integration of velocity versus time plot).

Example 2.1

Assume that a seismograph, located 1200 km from the epicenter of an earthquake, records a maximum ground displacement of 15.6 mm for surface waves having a period of 20 seconds. Based on these assumptions, determine the surface wave magnitude.

Solution:

Circumference of the earth $= 4.0 \times 10^7$ m (360°)

Distance to seismograph $= 1200$ km $= 1.2 \times 10^6$ m

$$\Delta = \frac{1.2\times10^6}{4\times10^7}(360^0) = 10.8^\circ$$

Using Eq. (2.2) and $A' = 15.6$ mm $= 15600$ μm gives:

$$M_s = \log A' + 1.66 \log \Delta + 2.0$$

$$= \log 15600 + 1.66 \log 10.8 + 2.0 = 7.9$$

Example 2.2

Assume that during a major earthquake, the depth of fault rupture is estimated to be 15 km, the length of surface faulting is determined to be 600 km, and the average slip along the fault is

2.5 m. Based on these assumptions, determine the moment magnitude. Use a shear modulus equal to 3 × 1010 N/m².

Solution: Use Eq. (2.3):

$$M_0 = \mu A_f D$$

where,

M_0 = seismic moment (N.m)

μ = shear modulus of material along fault plane

= 3×10^{10} N/m²

A_f = area of fault plane undergoing slip

= $15 \times 600 = 9000$ km² $= 9 \times 10^9$ m².

Δ = average displacement of ruptured segment of fault

= 2.5 m

Therefore,

$$M_0 = \mu A_f D = (3 \times 10^{10}\ \text{N/m}^2)(9 \times 10^9\ \text{m}^2)(2.5\ \text{m})$$

$$= 6.75 \times 10^{20}\ \text{N.m}$$

Using Eq. (2.4) gives:

$$M_w = -6.0 + 0.67 \log M_0 = -6.0 + 0.67 \log (6.75 \times 10^{20})$$

$$= 8$$

Home Work Problems

1. Assume that a seismograph, located 1000 km from the epicenter of an earthquake, records a maximum ground displacement of 15.6 mm for surface waves having a period of 20 seconds. Based on these assumptions, determine the surface wave magnitude. (**Ans.** $M_s = 7.777$)
2. Assume that during a major earthquake, the depth of full rupture is estimated to be 15 km, the length of surface faulting is determined to be 600 km, and the average slip along the fault is 2.5 m. Based on these assumptions, determine the moment magnitude. Use a shear modulus equal to 7×10^{12} N/m². (**Ans.** $M_w = 9.542$)
3. The standard Wood-Anderson seismograph has natural period of 0.8 sec, damping factor of 80% and static magnification of 2800. It is located exactly 100 km from the epicenter. Maximum trace amplitude recorded by it is 14.9 cm. Determine Richter magnitude. (**Ans.** $M_L = 5.2$)
4. Explain about main plate tectonic environments with example.
5. Present wave characteristics and particle motions for different main seismic waves in tabular form.
6. Write short note on fault structure.
7. Explain about dip, strike and slip.
8. Explain about dip-slip, strike-slip and oblique-slip type of hanging wall movement.
9. Present modified Mercalli scale in a tabular form.
10. Write short note about seismograph, seismogram and seismic waves obtained from seismogram.

SEISMIC HAZARDS IN INDIA

3.1 INTRODUCTION

Natural disasters like earthquake, landslide, flood, drought, cyclone, forest fire, volcanic eruption, epidemic and major accidents are quite common in different parts of the globe. These lead to the loss of life, property damage and socio-economic disruption. Such losses have grown over the years due to increase in population and physical resources. It is believed that the natural disasters have claimed more than 2.8 million lives during the past two decades only and have adversely affected 820 million people with a financial loss of about 25-100 million dollars. These losses are not evenly distributed and are more prevalent in the developing countries due to higher population concentration and low level of economic growth. United Nations in 1987 realized the need of reducing these losses due to natural disasters and proclaimed, by its Resolution No. 42/169, the current decade (1991-2000) as the International Decade for Natural Disaster Reduction (INDNDR). The main objective of this proclamation was to reduce, through concerted international efforts, the loss of life, property damage and socio-economic disruption caused by the natural disasters particularly in the developing countries.

Earthquakes are one of the worst among the natural disasters. About 1 lakh earthquakes of magnitude more than three hit the earth every year. According to a conservative estimate more than 15 million human lives have been lost and damage worth hundred billions of dollars has been inflicted in the recorded history due to these. Some of the catastrophic earthquakes of the world are Tangshan of China (1976, casualty > 3 lakhs), Mexico city (1985, casualty > 10,000) and North-West Turkey (August 17, 1999, casualty > 20,000). In India, casualty wise, the first three events are Kangra (>20,000), Bihar-Nepal (>10,653) and Killari (>10,000). Moreover, Indian-Subcontinent, particularly the northeastern region, is one of the most earthquakes-prone regions of the world.

Like any other natural disaster, it is not possible to prevent earthquakes from occurring. The disastrous effects of these, however, can be minimised considerably. This can be achieved through scientific understanding of their nature, causes, frequency, magnitude and areas of influence. The key word in this context is "Mitigation and Preparedness". Earthquake disaster mitigation and preparedness strategies are the need of the hour to fight and reduce its miseries to mankind. Comprehensive mitigation and preparedness planning includes avoiding hazard for instance. This can be achieved by providing warning to enable evacuation preceding the hazard, determining the location and nature of the earthquake hazard, identifying the population and structures vulnerable for hazards and adopting strategies to combat the menace of these. In the light of the above, the earthquake hazards in India have been discussed with special reference to the northeastern region along with the mitigation strategies.

3.2 EARTHQUAKE HAZARDS IN INDIA

Seismic zonation map shows that India is highly vulnerable for earthquake hazards. India has witnessed more than 650 earthquakes of Magnitude >5 during the last hundred years. Furthermore, the earthquake disaster is increasing alarmingly here. In addition to very active northern and northeastern seismicity, the recent events in Killari (Maharastra) and Jabalpur (Madhya Pradesh) in the Peninsular India have raised many problems to seismologists.

The occurrence of earthquakes can be explained with the concept of "Plate Tectonics". Based on this three broad categories of earthquakes can be recognised. (1) Those occurring at the subduction/collision zones. These are Inter-plates activity. (2) Those at mid-oceanic ridges and (3) Those at intra-plates (Acharrya, 1999). Seismic events in India mainly belong to the first category. However, a few third category events are also known. Earthquake events are reported from the Himalayan mountain range including Andaman and Nicobar Islands, Indo-Gangetic plain as well as from Peninsular region of India.

Subduction/collision earthquakes in India occur in the Himalayan Frontal Arc (HFA). This arc is about 2500 km long and extends from Kashmir in the west to Assam in the east. It constitutes the central part of the Alpine seismic belt. This is one of the most seismically active regions in the world. The Indian plate came into existence after initial rifting of the southern Gondwanaland in late Triassic period. Subsequently it drifted in mid-Jurassic to late Cretaceous time. The force responsible for this drifting came from the spreading of the Arabian Sea on either side of the Carisberg ridge. It eventually collided with the Eurasian plate. This led to the creation of Himalayan mountain range. The present day seismicity of this is due to continued collision between the Indian and the Eurasian plates. The important earthquakes that have visited HFA are tabulated below in Table 3.1.

Table 3.1 Important Earthquakes in HFA (Courtesy: http://gbpihed.nic.in)

Place	*Year*	*Magnitude*	*Casualty*
Kangra Valley	April 4, 1905	8.6	>20,000
Bihar-Nepal border	January 1, 1934	8.4	>10,653
Quetta	May 30, 1935	7.6	About 30,000
North Bihar	1988	6.5	1000 Approx.
Uttar Kashi	October 20, 1991	6.6	>2,000
Chamoli	March 29, 1999	6.8	>150
Hindukush	November 11, 1999	6.2	No death reported

The Peninsular India was once considered as a stable region. However, its seismic hazard status has increased due to the occurrence of damaging earthquakes (Pande, 1999). The recurrence intervals of these are, however, larger than those of the HFA. Furthermore, their magnitude is also lesser. These belong to intra-plate category of earthquakes. The important earthquake events that have rocked the Peninsular India is tabulated in Table 3.2.

Table 3.2 Important Earthquakes in Peninsular India (Courtesy: http://gbpihed.nic.in)

Place	*Year*	*Magnitude*	*Casualty*
Kutch	June 16, 1819	8.5	No record
Jabalpur	June 2, 1927	6.5	—
Indore	March 14, 1938	6.3	—
Bhadrachalam	April 14, 1969	6.0	—
Koyna	December 10, 1967	6.7	>200
Killari (Latur)	September 30, 1993	6.3	>10,000
Jabalpur	May 22, 1997	6.0	>55

Koyna event is a classic example of earthquake activity triggered by reservoir. Seismicity at Koyna has close correlation with the filling cycles of the Koyna reservoir. The most puzzling event in the Peninsular India is, however, the Killari earthquake. This occurred in the typical rural setting. This event was least expected from the tectonic consideration. Reason being, it is located in the Deccan Trap covered stable Indian shield. There is no record of any historical earthquake in this region. This has been considered as the most devastating SCR (Stable Continental Region) event in the world. Jabalpur event, which occurred in the urban centre, though moderate, is an important one. Reason being, it is the first major earthquake in India to be recorded by the newly established broadband digital station in the shield region. Furthermore, its spatial association with the Narmada Son lineament has triggered a lot of interest from the seismotectonic point of view (DST, 1999).

3.3 EARTHQUAKE HAZARDS IN THE NORTH EASTERN REGION

Northeastern region of India lies at the junction of the Himalayan arc to the north and the Burmese arc to the east. It is one of the six most seismically active regions of the world. The other five regions are Mexico, Japan, Taiwan, Turkey and California. Eighteen large earthquakes with magnitude >7 occurred in this region during the last hundred years (Kayal, 1998). High seismic activity in the northeastern region may be attributed to the collision tectonics in the north (Himalayan arc) and subduction tectonics in the east (Burmese arc). The Syntaxis Zone (The Mishmi Hills Block), which is the meeting place of the Himalayan and Burmese arcs is another tectonic domain in the region. The Main Central Thrust (MCT) and the Main Boundary Thrust (MBT) are the two major crystal discontinuities in the Himalayan arc of the Northeastern region. In the Burmese arc, the structural trend of the Indo-Myanmar Ranges (IMR) swing from the NE-SW in the Naga Hills to N-S along the Arakan Yoma and Chin Hills. Naga Thrust is the prominent discontinuity in the north. It connects the Tapu

Thrust to the south and Dauki Fault to the east. This fold belt appears to be continuous with the Andaman-Nicobar ridge to the south. The Mishmi Thrust and the Lohit Thrust are the major discontinuities identified in the Syntaxis Zone (Kayal, 1998). The list of important earthquake events in this region has been tabulated in Table 3.3.

Table 3.3 Important Earthquakes in Northeast India (Courtesy: http://gbpihed.nic.in)

Place	*Year*	*Magnitude*	*Remark*
Cachar	March 21, 1869	7.8	Numerous earth fissures and sand craters.
Shillong Plateau	June 12, 1897	8.7	About 1542 people died.
Sibsagar	August 31, 1906	7.0	Property damage.
Myanmar	December 12, 1908	7.5	Property damage.
Srimangal	July 8, 1918	7.6	4500 sq km area suffered damage.
SW Assam	September 9, 1923	7.1	Property damage.
Dhubri	July 2, 1930	7.1	Railway lines, culverts and bridges cracked.
Assam	January 27, 1931	7.6	Destruction of Property.
N-E Assam	October 23, 1943	7.2	Destruction of Property.
Upper Assam	July 29, 1949	7.6	Severe damage.
Upper Assam	August 15, 1950	8.7	About 1520 people died. One of the largest known quake in the history.
Indo-Myanmar border	August 6, 1988	7.5	No casualty reported.

The June 12, 1897 earthquake of the Shillong Plateau was one of the greatest event of the world. Casualty was only 1,542 compared to the magnitude of the event (8.7). This is so because event occurred at 5.15 p.m when most of the people were outdoor. Damage to the property was, however, severe. All concrete structures within an area of 30,000 square miles were practically destroyed. There was evidence of two surface faults, namely, Chidrang and Dudhnoi. It was the first instrumentally recorded event in the country. Another event of matching magnitude occurred on August 15, 1950 in the Syntaxis Zone. It caused 1520 death. However, it was more damaging than the 1897 event. Railway line and roads were considerably damaged. Landslide triggered in many places. Fissures and sand vents occurred. The last major event (magnitude = 7.5) in the region occurred on August 6, 1988. Its epicentre was in the Myanmar side of the IMR. This rocked the whole northeastern region. The tremor lasted for about two minutes, killing four human lives and damaging buildings, railway tracts and roads.

Fig. 3.1 at the end of chapter shows major tectonic features of the Indian Ocean showing spreading of Arabian Sea on either side of the Carlsberg Ridge.

3.4 FREQUENCY OF EARTHQUAKE

Seismologists seem not to believe that there is upheaval in the occurrence of earthquakes. Gupta (1999) says that annually on an average about 18 earthquakes of magnitude, which hit Turkey, (magnitude = 7.4), Greece (magnitude = 7.2) and Taiwan (magnitude = 7.6) recently occur all over the world. However, these were found to occur in uninhabited areas or virtually uninhabited areas. Unfortunately, these have now hit thickly populated areas. Consequently, they have killed thousands of people. This does not mean that the earthquake frequency has increased. Increase in the loss of life and property damage is due to increasing vulnerability of human civilization to these hazards. This can be understood by the fact that Kangra event of 1905 (magnitude = 8.6) and Bihar-Napal of 1934 (magnitude = 8.4) killed respectively about 20,000 and 10,653 people. On the other hand, 1897 and 1950 events of the northeast (magnitude = 8.7 each) could kill only about 1542 and 1520 people. This is because Kangra and Bihar-Nepal events struck densely populated areas of Indo-Gangetic plain. On the other hand, the northeastern region was sparsely populated in 1897 and 1950. Population concentration and physical resources have increased many times in this region since the last great event. Therefore, if the earthquake of matching magnitude visits the region now, the devastation would be enormous. Timing of the event and epicenter also matters a lot. For instance, Killari event occurred at 3:00 hrs early in the winter morning when people were sleeping and hence the casualty was high (>10,000). Similarly, Turkey event (August 17, 1999, magnitude = 7.4) also occurred 3.00 hrs in the morning when most people were asleep killing >20,000 of them. Jabalpur event, on the other hand, occurred 4.00 hrs in the morning on summer day when most of the people were outdoor. That is why, although the epicenter was near the town, the casualty was less (about 57). Property damage was also not that severe.

3.5 EARTHQUAKE PREDICTION

Research on earthquake prediction started since early sixties. Intensive work is going on all over the world in this regard involving expenditure of billions of dollars. The precise prediction of seismic events remains elusive and unattainable goal in spite of these efforts. According to R.R. Kelkar, Director General of Indian Meteorological Department (IMD), "Earthquake cannot be predicted by anyone, anywhere, in any country. This is a scientific truth". But seismologists continue their efforts in the hope of a major breakthrough in prediction technology in the near future. The seismologists are, however, in a position to indicate the possibility of recurrence of earthquakes in potentially large areas. This is based on palaeoseismicity, micro seismic activities as well as precursors.

It has been found that earthquakes are generally, preceded by some signals like ground tilting, foreshocks, change in ground water levels, variations in the discharge of springs, anomalous oil flow from the producing wells, enhance emanations of radon and unusual animal behaviour. However, this is not a sufficient condition. Perhaps the first successful prediction of earthquake in the world was made by the Chinese. They predicted Haicheng event of Lioning Province (February 4, 1975, magnitude = 7.3) on the basis of micro seismic activity, ground tilting and unusual animal behaviour (Nandi, 1999). They also foretold 4 out of 5 events of magnitude 7 during 1976-77. It is believed that the Chinese have mastered themselves in the art of closely monitoring and analysing animal behaviour to forecast earthquakes. Still they failed to predict Tangshan event of 1976 (magnitude = 7.8, casualty > 3 lakhs).

In India also efforts are going on for predicting earthquakes based on the statistical analysis of past events, their recurrence intervals, swarms activity and seismic gap. However, meaningful prediction is still alluding the seismologists. Khatri (1999) identified three seismic gaps in the Himalayan region. They are the Kashmir gap, the Central gap and the Assam gap. The Kashmir gap lies west of Kangra event, the Central gap between Kangra and Bihar-Nepal events and the Assam gap between the two great earthquakes of Assam. He further said that the great event may occur in these gaps in near future as well. Das and Sarmah (1996) has forecasted the occurrence of high magnitude earthquake in the western part of the northeastern region at any time within next few years. Negi (in Ahmad, 1998) has predicted "Mega Earthquake" in the northeast by 2010. The prediction has been made on the basis of theory of cyclical earthquakes. Sarmah (1999) calculated an average return period of 55 years for the earthquakes of magnitude 8 or greater. The last big earthquake of magnitude 8.7 occurred in 1950. Therefore, northeastern region is ready for an earthquake of similar magnitude. It is bare fact that the strain is accumulating in some parts of this region. Consequently, any delay in the occurrence of earthquake will increase its magnitude and thus the devastation only.

3.6 EARTHQUAKE HAZARD ZONATION, RISK EVALUATION AND MITIGATION

The importance of seismological studies lies in the fact that information generated can be used to mitigate the earthquake hazards. Preparation of seismotectonic/seismic zonation maps is the first step in this direction. The basic data required for the preparation of these maps are: (i) A carefully compiled earthquake catalogue incorporating details about magnitude, location of epicenter, depth of focus etc., (ii) Delineation of seismic source zones from all possible sources like recurrence relation, tectono-geological consideration, palaeoseismicity etc., (iii) Estimation of upper bound magnitude through statistical procedure, cumulative seismic energy release, active fault length *etc.* and (iv) Attenuation of ground shaking for better results (Das Gupta, 1999). Seismic microzonation is recommended for better result. These maps give an idea about the possibility of occurrence of earthquakes in the region and are very useful for evaluating the risk involved before designing and constructing the heavy engineering structures like dam, bridges, flyovers and large towers *etc.* These are also useful for planning human settlements that would remain safe during the occurrence of an earthquake. Seismic risk evaluation is also possible from these maps.

Indian Meteorological Department, National Geophysical Research Institute, Department of Science & Technology, Bhabha Atomic Research Centre and Regional Research Laboratory have established a large number of seismic monitoring network in the country including northeastern region. These stations are recording useful seismic data, which enables to determine the location of epicenter, depth of hypocenter, energy within the focus, orientation of the geological structure that has undergone deformation as well as many other parameters of earthquakes. These parameters are then utilised for preparing seismo-tectonic and seismic zoning maps. The work in seismic zoning in India was started by Indian Standard Institute (now Bureau of Indian Standard) in the year 1960. The first map was included in the code IS: 1893-1962. A significant progress has been made since then both in seismic zoning and instrumental monitoring of seismicity. However, many questions regarding the location and nature of potential seismic zones/faults still remain unsolved.

3.7 EARTHQUAKE RESISTANT STRUCTURES

It is necessary to design and construct earthquake resistant dwellings in the seismic prone zones. The principles of a seismic design should be kept in mind in this regard. The important earthquake resistant features which are recommended in the latest BIS codes (IS 13828:1993) should be followed (Bhagwan and Sreenath, 1996). Normally houses are built to withstand vertical load only. Consequently, they collapse when subjected to horizontal stresses produced by earthquake waves. The main requirements for preventing the collapse are a lateral load carrying system. The system should have enough residual capacity to safely resist lateral forces. It is said that the buildings made after 1981 basically had no damage due to Kobe event of 1995 in Japan because these fulfilled earthquake standards of construction (Struck, 1999). Besides, good quality construction materials should be used. The importance of quality material may be realised from the fact that almost all the individual houses in Jabalpur town withstood earthquake shaking. On the other hand, many government and private apartments built by contractors were badly damaged. It may be due to the reason that the contractors used substandard materials.

As northeastern region is highly seismic and experienced two great events of 1897 and 1950, the people here learnt to construct flexible and sufficiently earthquake proof houses. They are popularly known as "Assam Type" (Nandi, 1999). The scenario has changed now and these houses have paved the way for multistory masonry buildings. Most of them are in the capital towns of all the seven states of the region. If the present trend of construction and population growth continues, the earthquake of magnitude > 7.5 will bring enormous damage to property and great loss of lives. Consequently, the administrative agencies have to strictly enforce the implementation of proper building codes and appropriate land use policy in the region.

3.8 AWARENESS CAMPAIGN

Awareness campaign need to be launched to educate the people about the disastrous effects of earthquakes. Furthermore, people should be prepared to face them in a better way. Prevention and mitigation begins with the information. Moreover, public education and community participation is key to the success of the implementation of reduction and mitigation programmes.

A large number of specialised as well as popular articles have been written about earthquakes in research journals and conference proceedings, which are not available to common man. The newspapers and magazines usually do not show interest in publishing articles about mitigation and hazard reduction, however, they give extensive coverage after earthquake takes place. Information and popular articles should be written in simple language and be made readily available to common man. There has to be a close interaction between the seismologists and the administrators, which would greatly help the execution of seismic mitigation programmes (Bapat, 1996). Earthquake related curricula should be introduced in the school stage of education itself. Audio-visual programmes, preferably in the local languages have to be prepared and made available to the public. Voluntary organisation and college students may be approached to take up the responsibility of awareness campaign.

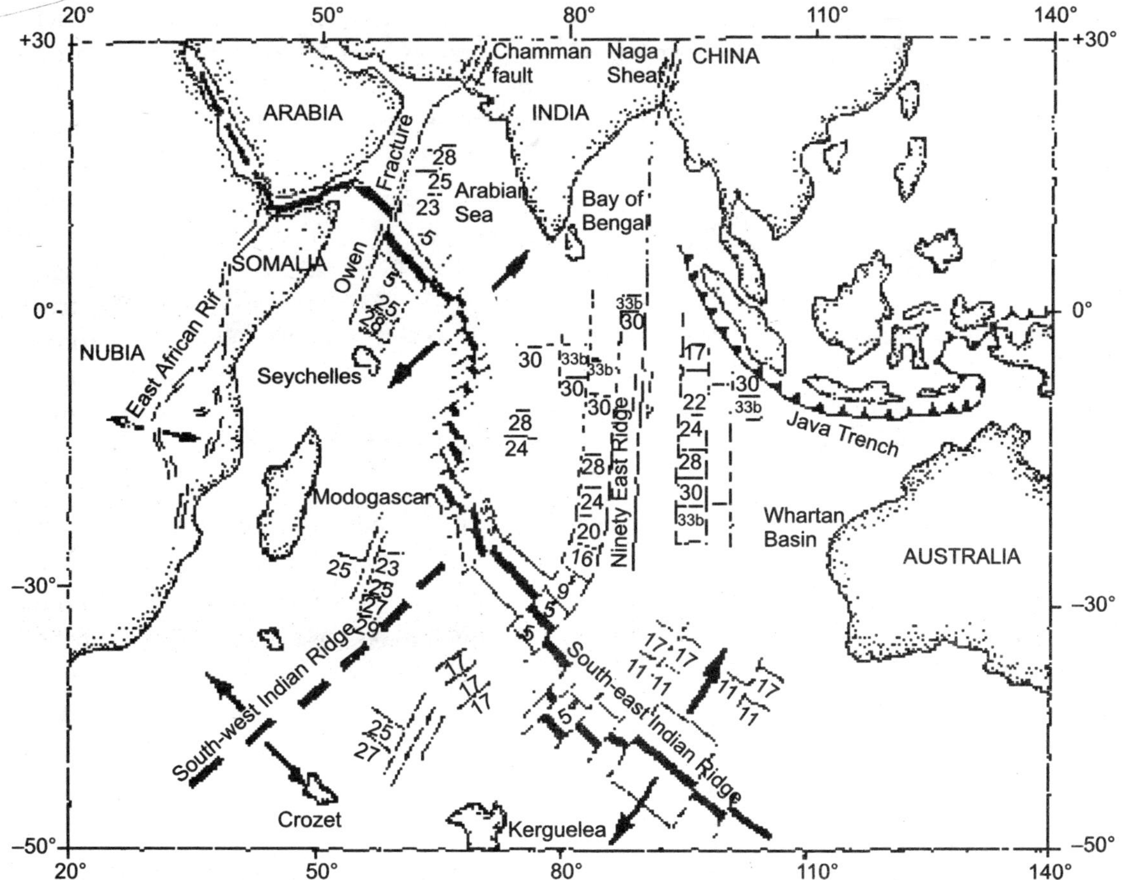

Fig. 3.1 Major tectonic features of the Indian Ocean showing spreading of Arabian Sea on either side of the Carlsberg Ridge (Courtesy: http://gbpihed.nic.in)

Home Work Problems

1. Explain occurrence of earthquakes in India with the concept of "Plate Tectonics".
2. Write short note on earthquake hazards in the Northeastern region of India.
3. Explain about earthquake resistant structures in India.

DYNAMIC SOIL PROPERTIES

4.1 INTRODUCTION

This book is concerned with geotechnical problems associated with dynamic loads. It also deals with earthquake related ground motion as well as soil response induced by earthquake loads. The dynamic response of foundations and structures depends on the magnitude, frequency, direction, and location of the dynamic loads. Furthermore, it also deals with the geometry of the soil-foundation contact system, as well as the dynamic properties of the supporting soils and structures.

Elements in a seismic response analysis are: input motions, site profile, static soil properties, dynamic soil properties, constitutive models of soil response to loading and methods of analysis using computer programs. The contents include: earthquake response spectra; site seismicity; soil response to seismic motion, design earthquake, seismic loads on structures, liquefaction potential, lateral spread from liquefaction, and foundation base isolation.

Some special problems in geotechnical engineering dealing with soil dynamics and earthquake aspects are discussed in the later chapters. Its contents include: liquefaction potential of soil, foundation settlement, dynamic bearing capacity of foundations, stone columns and displacement piles, dynamic slope stability and dynamic earth pressure in the context of earthquake loading.

4.2 SOIL PROPERTIES FOR DYNAMIC LOADING

The properties that are most important for dynamic analyses are the stiffness, material damping, and unit weight. These properties enter directly into the computations of dynamic response. In addition, the location of the water table, degree of saturation, and grain size distribution may be important, especially when liquefaction is a potential problem. Since earthquake induces dynamic kind of loading into the soil, these dynamic soil properties are quite significant.

One method of direct determination of dynamic soil properties in the field is to measure the velocity of shear waves in the soil. The waves are generated by impacts produced by a hammer or by detonating charges of explosives. Then the Travel times are recorded. This is usually done in or between bore holes. A rough correlation between the number of blows per foot in standard penetration tests and the velocity of shear waves is shown in Fig. 4.1. Standard penetration test is well known test in foundation engineering.

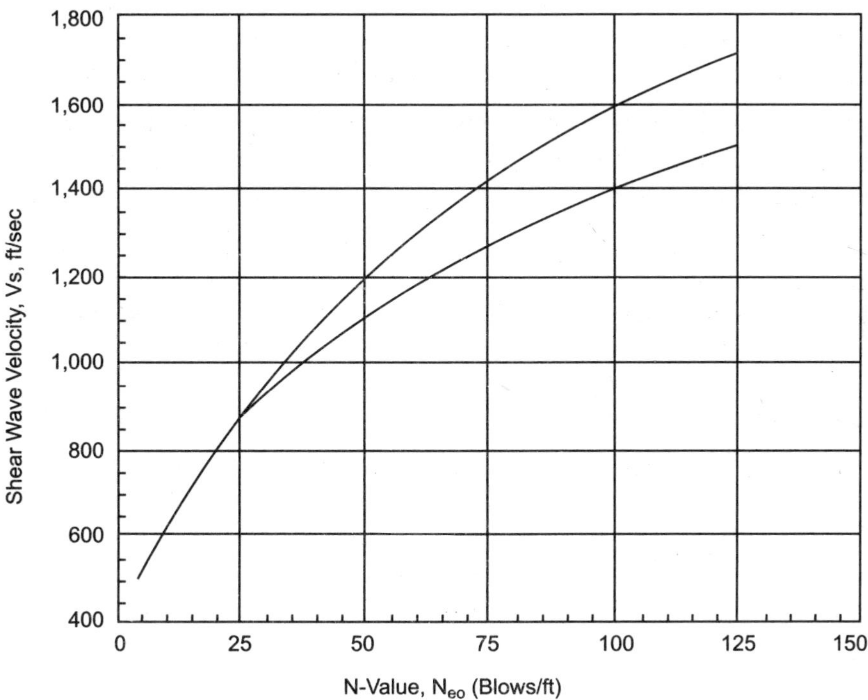

Fig. 4.1 Relation between number of blows per foot in standard penetration test and velocity of shear waves (Courtesy: <http://www.vulcanhammer.net>)

4.3 TYPES OF SOILS

As in other areas of soil mechanics, the type of the soil affects its response under dynamic loading conditions. Furthermore, it also determines the type of dynamic problems that must be analyzed. The most significant factors separating different types of soils is the grain size distribution. The presence or absence of clay fraction in soil system, as well as the degree of saturation of soil system also plays key role in this connection. It is also important to know whether the dynamic loading is a transient phenomenon, such as a blast loading or earthquake, or is a long term phenomenon, like a vibratory loading from rotating machinery. The distinction is important because a transient dynamic phenomenon occurs so rapidly that excess pore pressure does not have time to dissipate. Dissipation of pore water is possible only in the case of very coarse, clean gravels if dynamic loading is a transient dynamic phenomenon. In this context the length of the drainage path is also important. Even a clean, granular material may retain large excess pore pressure if the drainage path is so long that the pressures cannot dissipate during the dynamic loading. Consequently, it is necessary to categorize the soil by asking the following questions:

(*a*) Is the material saturated? If it is saturated, a transient dynamic loading will usually last for very short duration. The duration is so short that the soil's response is essentially undrained. If it is not saturated, the response to dynamic loadings will probably include some volumetric component as well.

(*b*) Are there fines present in the soil? The presence of fines, especially clays inhibits the dissipation of excess pore pressure. It also decreases the tendency for liquefaction.

(*c*) How dense is the soil? Dense soils are not likely to collapse under dynamic loads. On the other hand, Loose soils may collapse under dynamic loads. Furthermore, Loose soils may densify under vibratory loading and cause permanent settlements.

(*d*) How are the grain sizes distributed? Well graded materials are less susceptible to losing strength under dynamic loading. On the other hand Uniform soils are more susceptible to losing strength under dynamic loading. Loose, Uniform soils are especially subject to collapse and failure under dynamic loading.

4.3.1 Dry and Partially Saturated Cohesionless Soils

There are three types of dry or partially saturated Cohesionless Soils The first type comprises soils that consist essentially of small-sized to medium-sized grains of sufficient strength or under sufficiently small stress condition. The grain breakage does not play a significant role in their behavior. The second type includes those soils made up essentially of large-sized grains, such as rockfills. Large-sized grains may break under large stresses. Overall volume changes are significantly conditioned by grain breakage. The third type includes fine-grained materials, such as silt. The behavior of the first type of dry cohesionless soils can be described in terms of the critical void ratio. The behavior of the second type depends on the normal stresses and grain size. If the water or air cannot escape at a sufficiently fast rate when the third type of soil is contracting due to vibration under dynamic loading, significant pore pressures may develop. Consequently liquefaction of the material is likely.

4.3.2 Saturated Cohesionless Soils

If pore water can flow in and out of the material at a sufficiently high rate, pore pressures do not develop. Consequently, behavior of these soils does not differ qualitatively from that of partially saturated cohesionless soils. If the pore water cannot flow in or out of the material, cyclic loads under dynamic load will usually generate increased pore pressure. If the soil is loose or contractive, the soil may liquefy.

4.3.3 Saturated Cohesive Soils

Alternating loads decrease the strength and stiffness of cohesive soils. The decrease depends on the number of repetitions. It also depends on the relative values of sustained and cycling stresses as well as on the sensitivity of the soil. Very sensitive clays may lose so much of their strength that there may be a sudden failure. The phenomenon is associated with a reduction in effective pressure as was the case with cohesionless soils.

4.3.4 Partially Saturated Cohesive Soils

The discussion in connection with saturated Cohesive soils, are applied to insensitive soils as well, when they are partially saturated, except that the possibility of liquefaction seems remote in the later kind of soil.

4.4 MEASURING DYNAMIC SOIL PROPERTIES

Soil properties to be used in dynamic analyses can be measured in the field. These properties can also be measured in the laboratory. In many important applications, a combination of field and laboratory measurements are used.

4.4.1 Field Measurements of Dynamic Modulus

Direct measurement for soil or rock stiffness in the field has the advantage of minimal material disturbance. The modulus is measured where the soil exists. Furthermore, the measurements are not constrained by the size of a sample.

Moduli measured in the field correspond to very small strains. Some procedures for measuring moduli at large strain have also been proposed. However, none has been found fully satisfactory by the geotechnical engineering community. The dissipation of energy during strain, which is called material damping, requires significant strains to occur. Consequently, field techniques have failed to prove effective in measuring material damping.

In situ techniques are based on measurement of the velocity of propagation of stress waves through the soil. The P-waves or compression waves are dominated by the response of the pore fluid in the saturated soils. Consequently, most techniques measure the S-waves or shear waves. If the velocity of the shear wave through a soil deposit is determined to be V_s, the shear modulus G is given as:

$$G = \rho V_s^2 = \frac{\gamma}{g} V_s^2 \tag{4.1}$$

where, ρ = mass density of soil.

γ = unit weight of soil.

g = acceleration of gravity.

There are three techniques for measuring shear wave velocity in in-situ soil. These techniques are as follows: cross-hole, down-hole, and uphole. All the three techniques require boring to be made in the in-situ soil.

In the cross-hole method sensors are placed at one elevation in one or more borings. Then a source of energy is triggered in another boring at the same elevation. The waves travel horizontally from the source to the receiving holes. The arrivals of the S-waves are noted on the traces of the response of the sensors. The velocity of S-wave can be calculated by dividing the distance between borings by the time for a wave to travel between them. However, it is difficult to establish the exact triggering time. Consequently, the most accurate measurements are obtained from the difference of arrival times at two or more receiving holes rather than from the time between the triggering and the arrival at single hole.

P-waves travel faster than S-waves. Consequently, the sensors will already be excited by the P-waves when the S-waves arrive. This can make it difficult to pick out the arrival of the S-wave. To alleviate this difficulty it is desirable to use an energy source that is rich in the vertical shear component of motion and relatively poor in compressive motion. Several devices are available that do this. The original cross-hole velocity measurement methods used explosives as the source of energy. These were rich in compression energy and poor in shear energy. Consequently, it is quite difficult to pick out the S-wave arrivals in this case. Hence, explosives should not be used as energy sources for cross-hole S-wave velocity measurements. ASTM D 4428/D 4428M, Cross-Hole Seismic Testing, describes the details of this test.

In the down-hole method the sensors are placed at various depths in the boring. Furthermore, the source of energy is above the sensors - usually at the surface. A source rich in S-waves should be used. This technique does not require as many borings as the cross-hole method. However, the waves travel through several layers from the source to the sensors. Thus, the measured travel time reflects the cumulative travel through layers with different wave velocities. Interpreting the data requires sorting out the contribution of the layers. The seismocone version of the cone penetration test is one example of the down-hole method.

In the up-hole method the source of the energy is deep in the boring. The sensors are above it—usually at the surface.

A recently developed technique that does not require borings is the spectral analysis of surface waves (SASW). This technique uses sensors that are spread out along a line at the surface. The source of energy is a hammer or tamper also located at the surface. The surface excitation generates surface waves. In particular, they are Rayleigh waves. These are waves that occur because of the difference in stiffness between the soil and the overlying air. The particles move in retrograde ellipses and their amplitudes decay from the surface. The test results are interpreted by recording the signals at each of the receiving stations. Computer program is used to perform the spectral analysis of the data. Computer programs have been developed that will determine the shear wave velocities from the results of the spectral analysis.

The SASW method is most effective for determining properties near the surface. In order to increase the depth of the measurements, the energy at the source must also be increased. Measurements for the few feet below the surface, which may be adequate for evaluating pavements, can be accomplished with a sledge hammer as a source of energy. However, measurements several tens of feet deep require track-mounted seismic "pingers." The SASW method works best in cases where the stiffness of the soils and rocks increases with depth. If there are soft layers lying under stiff ones, the interpretation may be ambiguous. A soft layer lying between stiff ones can cause problems for the crosshole method as well. Reason being that the waves will travel fastest through the stiff layers and the soft layer may be masked.

The cross-hole, down-hole, and up-hole methods may not work well very near the surface. Complications due to surface effects may affect the readings while using aforementioned methods. This is the region where the SASW method should provide the best result. The crosshole technique employs waves with horizontal particle motion. The down-hole and up-hole methods use waves whose particle motions are vertical or nearly so. Surface waves in the SASW method have particle motions in all the sensors. Therefore, a combination of these techniques can be expected to give a more reliable picture of the shear modulus than any one used alone.

4.4.2 Laboratory Measurement of Dynamic Soil Properties

Laboratory measurements of soil properties can be used to supplement or confirm the results of field measurements. They can also be necessary to establish values of damping and modulus at strains larger than those that can be attained in the field. Furthermore, they are also used to measure the properties of materials that do not presently exist in the field. Example is soil to be compacted.

A large number of laboratory tests for dynamic purposes have been developed. Research is continuing in this area. These tests can generally be classified into two groups. First group of tests are those that apply dynamic loads. Second group of tests are those that apply loads that are cyclic but slow enough that inertial effects do not occur.

The most widely used of the laboratory tests that apply dynamic loads is the resonant-column method. In this test a column of soil is subjected to an oscillating longitudinal or torsional load. The frequency is varied until resonance occur. From the frequency and amplitude at resonance the modulus and damping of the soil can be calculated. A further measure of the damping can be obtained by observing the decay of oscillations when the load is cut off.

ASTM D 4015 describes only one type of resonant-column device. However, there are several types that have been developed. These devices provide measurements of both modulus and damping at low strain levels. The strains can sometimes be raised a few percent. However, they remain essentially low strain devices. These devices could be of torsional or of longitudinal type. The torsional devices give measurements on shear behavior. On the other hand, the longitudinal devices give measurements pertaining to extension and compression behavior.

The most widely used of the cyclic loading laboratory tests is the cyclic triaxial test. In this test a cyclic load is applied to a column of soil over a number of cycles. Cyclic load application is slow, such that inertial effects do not occur. The response at one amplitude of load is observed. Afterwards, the test is repeated at a higher load. Fig. 4.2(A) shows the typical pattern of stress and strain. It is expressed as shear stress and shear strain. The shear modulus is the slope of the secant line inside the loop in Fig. 4.2(A). The critical damping ratio, D, is:

$$D = \frac{A_i}{4\pi A_T} \quad \text{...(4.2)}$$

where,

A_i = area of loop

A_t = shaded area

Other types of cyclic loading devices also exist. Cyclic simple shear devices are such devices. Their results are interpreted similarly. These devices load the sample to levels of strain much larger than those attainable in the resonant column devices. A major problem in both resonant-column and cyclic devices is the difficulty of obtaining undisturbed samples. This is especially true for small-strain data. Reason being that the effects of sample disturbance are particularly apparent at small strains.

The results of laboratory tests are often presented in a form similar to Fig. 4.2 (B-1 and B-2). In Fig. 4.2 (B-1) the ordinate is the secant modulus divided by the modulus at small strains. In Fig. 4.2 (B-2) the ordinate is the value of the initial damping ratio. Both are plotted against the logarithm of the cyclic strain level.

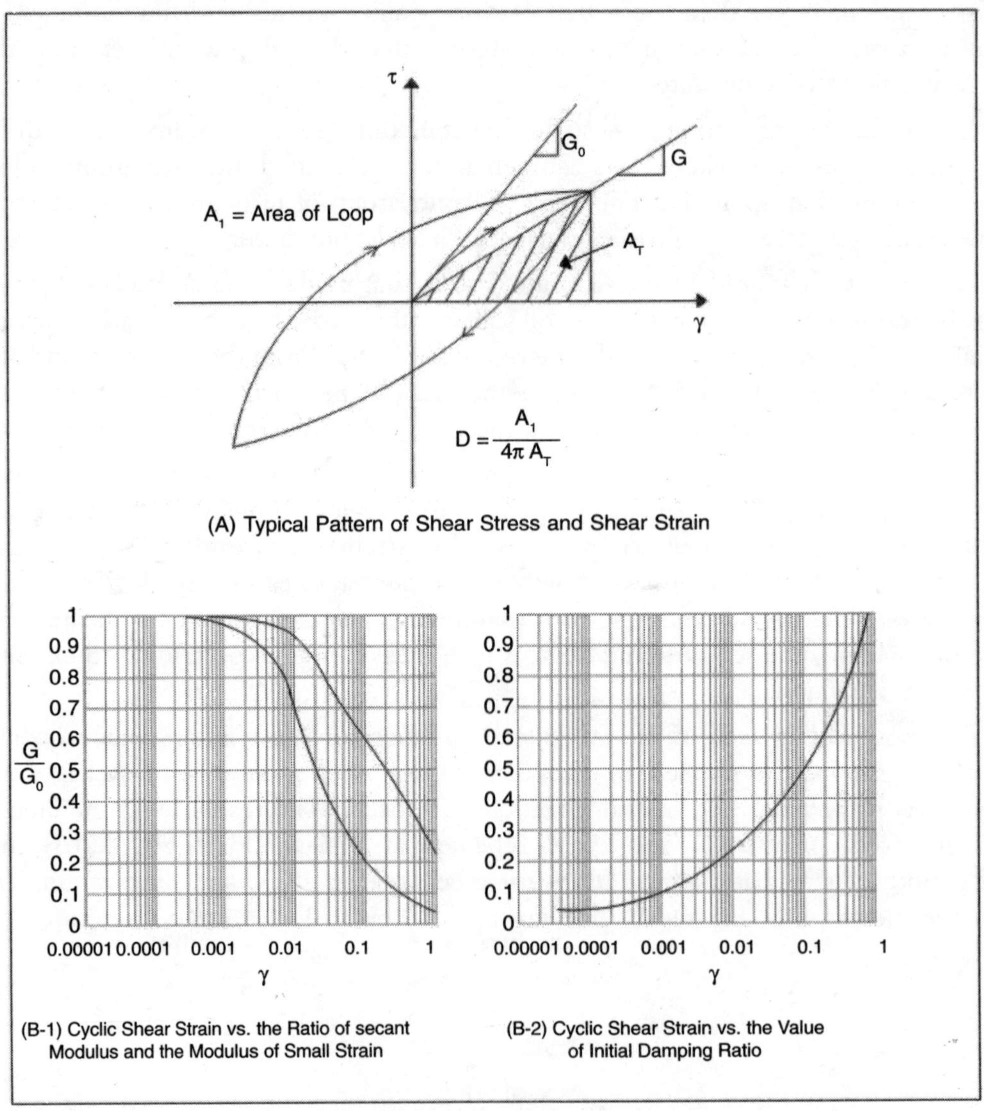

Fig. 4.2 Laboratory measurement of dynamic soil properties
(Courtesy: <http://www.vulcanhammer.net>)

Home Work Problems

1. The shear moduli of steel and its specific gravity is 11.5×10^6 psi and 7.85 respectively. Determine shear wave velocity through it. (**Ans.** 10434 ft/sec)
2. The type of the soil affects its response under dynamic loading conditions. Justify the statement.
3. Explain about standard techniques for measuring shear wave velocity in in-situ soil.
4. Explain about cyclic triaxial testing. How shear modulus and critical damping ratio is determined using cyclic triaxial testing.

SITE SEISMICITY, SEISMIC SOIL RESPONSE AND DESIGN EARTHQUAKE

5.1 SITE SEISMICITY

5.1.1 Site Seismicity Study

The objective of a seismicity study is to quantify the level and characteristics of ground motion shaking associated with earthquake that pose a risk to a given site of interest. A seismicity study starts with detailed examination of available geological, historical, and seismological data. These data are used to establish patterns of seismicity. They are also used to locate possible sources of earthquakes and their associated mechanisms. The site seismicity study produces a description of the earthquake for which facilities must be designed. In many cases, this will take the form of a probability distribution of expected site acceleration (or other measurements of ground motion) for a given exposure period. It will also give an indication of the frequency content of that motion. In some cases typical ground motion time histories called scenario earthquakes are developed. One approach is to use the historical epicenter database in conjunction with available geological data. These data are used to form a best estimate regarding the probability of site ground motion.

Fig. 5.1 explains some terms that are commonly used in seismic hazard analysis. The "hypocenter" or "focus" is the point at which the motions originated. This is usually the point on the causative fault. This is the point at which the first sliding occurs. It is not necessarily the point from which greatest energy is propagated. The "epicenter" is the point on the ground surface that lies directly above the focus. The "focal depth" is the depth of the focus below the ground surface. The "epicentral distance" is the distance from the epicenter to the point of interest on the surface of the earth. These aspects have been discussed in Chapter 2 also.

As a part of the Navy's seismic hazard mitigation program, procedures were developed in the form of a computer program (named SEISMIC, NAVFACENGCOM technical report

TR-2016-SHR, procedures for computing site seismicity, and acceleration in rock for earthquakes in the western united states). The program was designed to run on standard desktop DOS-based computers. The procedures consist of:

(*a*) Evaluating tectonics and geologic settings.

(*b*) Specifying faulting sources.

(*c*) Determining site soil conditions.

(*d*) Determining the geologic slip rate data.

(*e*) Specifying the epicenter search area and search of database.

(*f*) Specifying and formulating the site seismicity model.

(*g*) Developing the recurrence model.

(*h*) Determining the maximum source events.

(*i*) Selecting the motion attenuation relationship.

(*j*) Computing individual fault/source seismic, contributions.

(*k*) Summing the effects of the sources.

(*l*) Determining the site matched spectra for causative events.

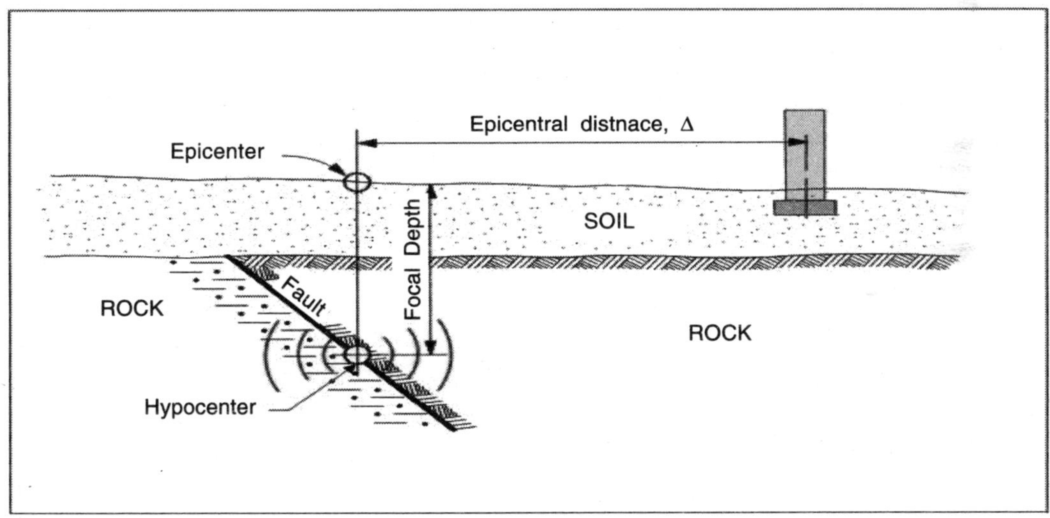

Fig. 5.1 Definition of earthquake terms (Courtesy: <http://www.vulcanhammer.net>)

5.1.2 Ground Motion Estimates

Ground motion attenuation equations are used to determine the level of acceleration as a function of distance from the source as well as the magnitude of the earthquake. Correlations have been made between peak acceleration and other descriptions of ground motion with distance for various events. These equations allow the engineers to estimate the ground motions at a site from a specified event. They also allow engineers to find out the uncertainty associated with the estimate. There are a number of attenuation equations that have been developed by various researchers. Donovan and Bornstein, 1978, developed the following equation for peak horizontal acceleration. Equations were developed from the western united states data.

$$Y = (a)(\exp(bM))(r + 25)^{d} \quad ...(5.1a)$$

$$a = (2{,}154{,}000)(r)^{-2.10} \quad ...(5.1b)$$

$$b = (0.046)+(0.445)\log(r) \quad ...(5.1c)$$

$$d = (2.515)+(0.486)\log(r) \quad ...(5.1d)$$

where, Y = peak horizontal acceleration (in gal) (1 gal = 1 cm/sec^2)

M = earthquake magnitude

r = distance (in km) to energy center, default at a depth of 5 km.

5.1.3 Analysis Techniques

NAVFAC P355.1, seismic design guidelines for essential buildings provides instructions for site seismicity studies. These studies are used for determining ground motion and response spectra. An automated procedure has been developed by NFESC (Naval Facilities Engineering Service Center) to perform a seismic analysis. The analysis has been done using available historic and geological data to compute the probability of occurrence of acceleration at a given site. A regional study is first performed in which all of the historic epicenters are used with an attenuation relationship. This study is used to compute the site acceleration for all historic earthquakes. A regression analysis is performed to obtain regional recurrence coefficients, and a map of epicenters is plotted. Confidence bounds are given on the site acceleration as a function of probability of exceedance.

5.2 SEISMIC SOIL RESPONSE

5.2.1 Seismic Response of Horizontally Layered Soil Deposits

Several methods for evaluating the effect of local soil conditions on ground response during earthquakes are now available. Most of these methods are based on the assumption that the main responses in a soil deposit are caused by the upward propagation of horizontally polarized shear waves (SH waves). These waves are propagated from the underlying rock formation. Analytical procedures based on this concept incorporating linear approximation to nonlinear soil behavior, have been shown to give results in fair agreement with field observations in a number of cases. Accordingly, engineers are finding increasing use in earthquake engineering for predicting response within soil deposits and the characteristics of ground surface motions.

5.2.2 Evaluation Procedure

The analytical procedure generally involves the following steps:

(*a*) Determine the characteristics of the motions likely to develop in the rock formation underlying the site. After that select an accelerogram with these characteristics for use in the analysis. The maximum acceleration, predominant period, and effective duration are the most important parameters of an earthquake motion. Empirical relationships between these parameters and the distance from the causative fault to the site have been established for earthquakes of different magnitudes. A design motion with the desired characteristics can be selected from the strong motion accelerograms that have been recorded during previous earthquakes or from artificially generated accelerograms.

(*b*) Determine the dynamic properties of the soil deposit. Average relationships between the dynamic shear moduli, as functions of shear strain and static properties, have been established for various soil types (Seed and Idriss, 1970). Average relation between the damping ratios of soils, as functions of shear strain and static properties have also been established. Thus a testing program to obtain the static properties for use in these relationships will often serve to establish the dynamic properties with a sufficient degree of accuracy. However more elaborate dynamic testing procedures are required for special problems. These techniques are also needed for soil types for which empirical relationships with static properties have not been established.

(*c*) Compute the response of the soil deposit to the base rock motions. A one-dimensional method of analysis can be used if the soil structure is essentially horizontal. Computer programs developed for performing this analysis are generally based on either the solution to the wave equation or on a lumped mass simulation. More irregular soil deposits may require a finite element analysis.

5.2.3 Analysis Using Computer Program

A computer program SHAKE, which is based on the one dimensional wave propagation method is available. The program can compute the responses for a design motion given anywhere in the system. Thus acceleration obtained from instruments on soil deposits can be used to generate new rock motions which, in turn, can be used as design motion for other soil deposits. Fig. 5.2 shows schematic representation of the procedure for computing effects of local soil conditions on ground motions. If the ground motions are known or specified at Point A, the SHAKE program can be used to compute the motion to the base of the soil column. That is, the program finds the base rock motion that causes the motion at Point A. The program can then find what the motion would be at a rock outcrop if the base rock motion had been propagated upward through rock instead of soil. This rock outcrop motion is then used as input to an amplification analysis, yielding the motion at Point B. From Fig. 5.2 it is clear that it is the top of another soil column.

The program also incorporates a linear approximation to nonlinear soil behavior. Furthermore, it also incorporates the effect of the elasticity of the base rock, and systems with different values of damping and modulus in different layers. Other versions of the same sort of analysis, often incorporating other useful features, are also available and may be superior to the original version of SHAKE. A NAVFAC sponsored MSHAKE microcomputer program was developed in 1994. The MSHAKE is a user friendly implementation of the SHAKE91 program which is a modified version of the original computer program SHAKE.

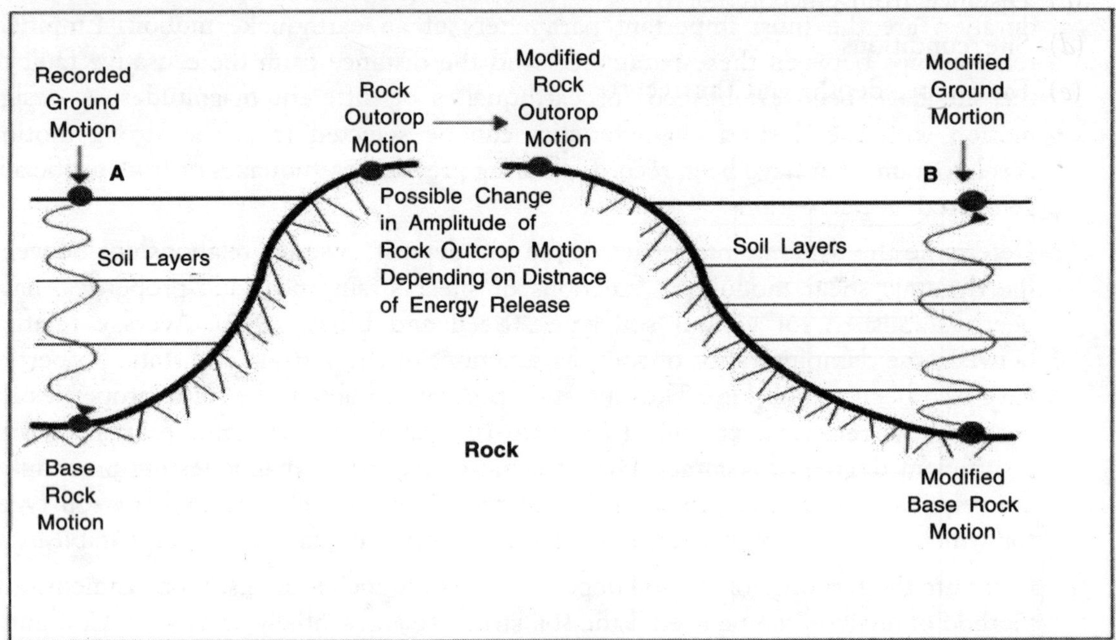

Fig. 5.2 Schematic representation of procedure for computing effects of local soil conditions on ground motions (Courtesy: <http://www.vulcanhammer.net>)

5.3 DESIGN EARTHQUAKE

5.3.1 Design Parameters

In evaluating the soil behavior under earthquake motion, it is necessary to know the magnitude of the earthquake. It is also necessary to describe the ground motion in terms that can be used for further engineering analysis. Historically, design earthquake waves were specified in terms of the peak acceleration. However, more modern techniques use the response spectrum or one or more time histories of motion. It has been concluded that the most reliable method for accomplishing this is to base the studies on data obtained at the site. A second choice is to find another site similar in geologic and seismic setting where ground motion was measured during a design level magnitude earthquake. However, this will usually not be possible, and estimates of ground motion based on correlations and geologic and seismologic evidence for the specific site will become necessary.

Factors Affecting Ground Motion. Factors that affect strong ground motion include:

(*a*) Wave types—S and P waves that travel through the earth, as well as the surface waves that propagate along the surfaces or interfaces.

(*b*) Earthquake magnitude—There are several magnitude scales. Even a small magnitude event may produce large accelerations in the near field. Consequently, a wide variety of acceleration for the same magnitude may be expected.

(*c*) Distance from epicenter or from center of energy release.

(*d*) Site conditions.

(*e*) Fault type, depth, and the recurrence interval.

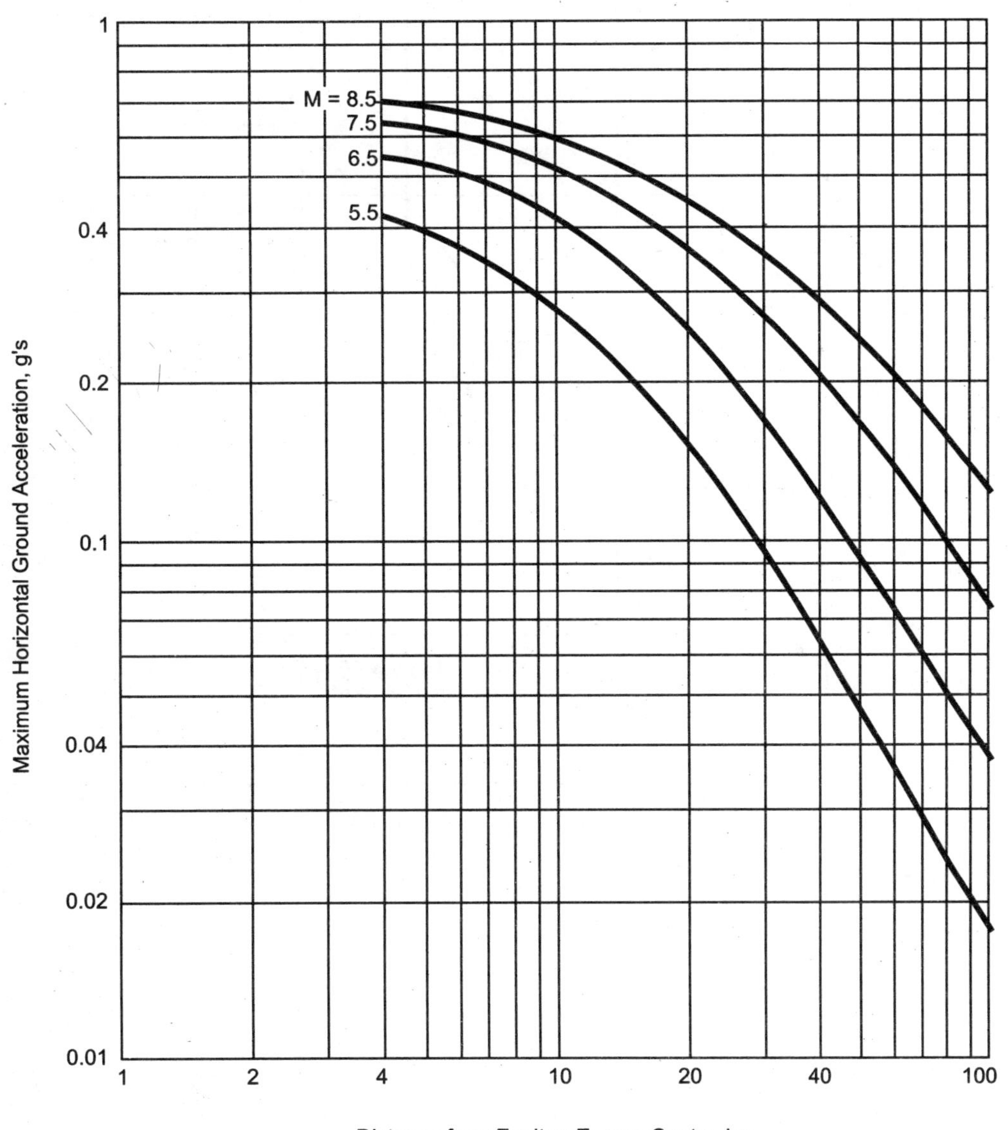

Fig. 5.3 Example of attenuation relationships in rock (Courtesy: http://www.valcanhammer.net)

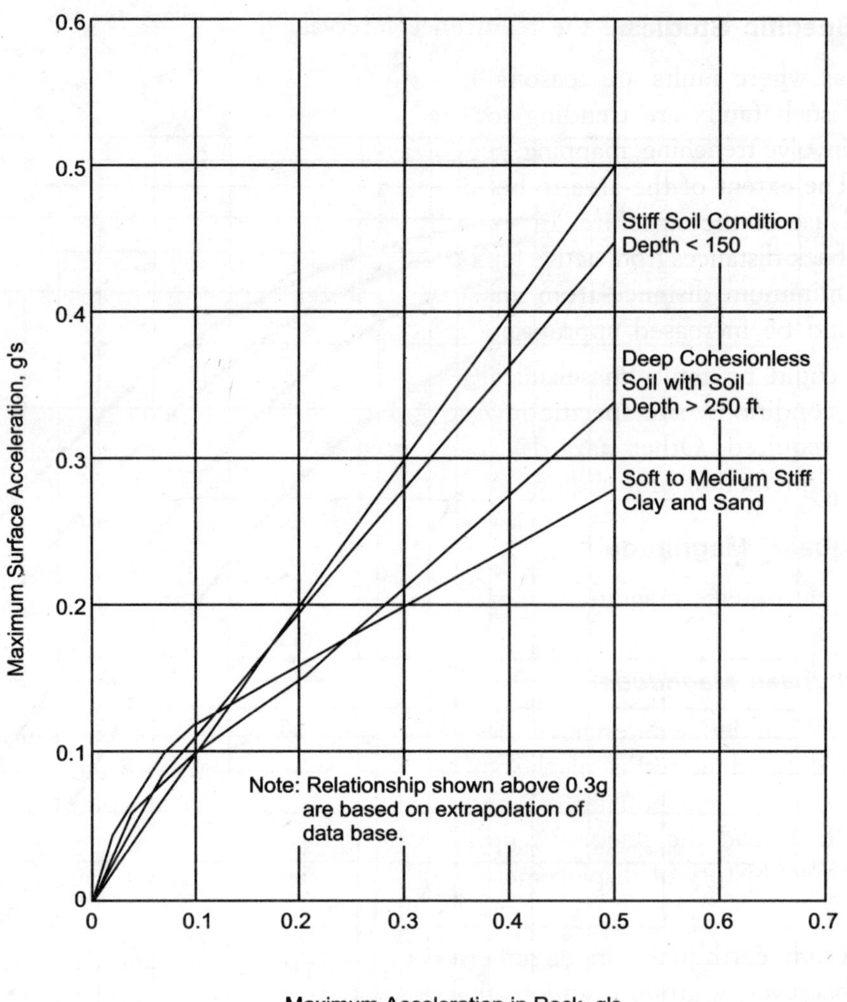

Fig. 5.4 approximate relationship for maximum acceleration in various soil conditions knowing maximum acceleration in rock (Courtesy: http://www.vulcanhammer.net)

Ground Motion Parameters

Ground motion parameters have been correlated with magnitude and distance. These correlations have been developed by several investigators. The correlation in Fig. 5.3 (Schnabel and Seed, 1973), is based on ground motion records from the Western United States. Furthermore, it is believed to be more applicable to small and moderate earthquakes (magnitudes 5.5 and 6.5) for rock. This correlation is also statistically applicable for stiff soil sites (e.g., where overburden is of stiff clays and dense sands less than 150 feet thick). For other site conditions, motion may occur as illustrated in Fig. 5.4 (Relationship between maximum acceleration, maximum velocity, distance from source and local site conditions for moderately strong earthquake, seed, murnaka, lysmer, and idris, 1975).

5.3.2 Site Specific Studies

In areas where faults are reasonably mapped and studied, site specific investigations can verify if such faults are trending towards the site or the facility is on an active fault. Studies may involve trenching, mapping, geophysical measurements, as well as other investigation techniques. The extent of the area to be investigated depends on geology. It also depends on the type and use of the structure. In some localities, state, or local building codes establish minimum setback distances from active faults. Unless other critical conditions demand differently, 300 feet of minimum distance from an active fault is provided. For essential facilities the distance should be increased appropriately.

There could be faults in seismically active areas where faults are not well mapped. Under these conditions, site specific investigations may be required. Regional investigations may also be required. Other hazards to be considered in a site investigation include the potential for liquefaction and sliding.

5.3.3 Earthquake Magnitude

Design earthquake magnitude as well as the selection of magnitude level are discussed below:

Design Earthquake Magnitude

Engineers can define a design earthquake for a site in terms of the earthquake magnitude, M. It is also defined in terms of the strength of ground motion. Factors influencing the selection of a design earthquake are the length of geologic fault structures, relationship between the fault and the regional tectonic structure, the rate of displacement across the fault, the geologic history of displacement along the structure, and the seismic history of the region.

The design earthquake in engineering terms is a specification of levels of ground motion. At this level of ground motion, the structure is required to survive successfully with no loss of life, acceptable damage, or no loss of service. A design earthquake on a statistical basis considers the probability of the recurrence of a historical event.

Earthquake magnitudes can be specified in terms of a design level earthquake. This level of earthquake can reasonably be expected to occur during the life of the structure. As such, this represents a service load that the structure must withstand without significant structural damage or interruption of a required operation. A second level of earthquake magnitude is a maximum credible event for which the structure must not collapse. However, significant structural damage can occur. The inelastic behavior of the structure must be limited to ensure the prevention of collapse and catastrophic loss of life during earthquake.

Selection of Design Earthquake

The selection of a design earthquake may be based on:

(*a*) Known design-level and maximum credible earthquake magnitudes. These are associated with a fault whose seismicity has been estimated.

(*b*) Specification of probability of occurrence of the earthquake for a given life of the structure (such as having a 10 percent chance of being exceeded in 50 years).

(*c*) Specification of a required level of ground motion. This specification is available in the code provision.

(*d*) Fault length empirical relationships.

The original magnitude scale proposed by Gutenburg and Richter is calculated from a standard earthquake. The standard earthquake is the one which provides a maximum trace amplitude of 1 mm on a standard Wood Anderson torsion seismograph at a distance of 100 km. Magnitude is the log10 of the ratio of the amplitude of any earthquake at the standard distance to that of the standard earthquake. Each full integer step in the scale (two to three, for example) represents an energy increase of about 32 times.

Because of the history of seismology, there are actually several magnitude scales. Modern earthquakes are described by the moment magnitude, Mw. Earlier earthquake events may be described by any of a number of other scales. Fortunately, the numerical values are usually within 0.2 to 0.3 magnitude units for magnitudes up to about 7.5. For larger events the values deviate significantly.

5.3.4 Intensity

In areas where instrumental records are not available the strength of an earthquake has usually been estimated on the basis of the modified Mercalli (MM) intensity scale. The MM scale is a number based mostly on subjective description of the effects of earthquakes on structures and people. Intensity is a very qualitative measure of local effects of an earthquake. Magnitude is a quantitative measure of the size of the earthquake at its source. The MM intensity scale has been correlated with peak horizontal ground acceleration by several investigators. It has been illustrated in Fig. 5.5.

5.3.5 Peak Horizontal Ground Acceleration

Seismic investigations of activities located in different seismic zones have been made. These seismic investigations include a site seismicity study. Where such studies have been completed, they help in determining the peak horizontal ground acceleration. Where a site seismicity study has not yet been completed, it may be warranted in connection with the design and construction of an important new facility. In connection with soil related calculations, the peak horizontal ground acceleration for Seismic Zone 2 may be taken as 0.17g and for Seismic Zone 1 as 0.1g. 'g' refers to acceleration due to gravity. There are standard techniques of making seismic zones.

5.3.6 Seismic Coefficients

The values of seismic ground acceleration can be determined from Table 5.1 based on soil types defined as follows:

(*a*) Rock with 2500 ft/sec < V_s < 5000 ft/sec (760 m/sec < V_s < 1520 m/sec)

(*b*) Very dense soil and soft rock with 1200 ft/sec < V_s < 2500 ft/sec (360 m/sec < Vs < 760 m/sec)

(*c*) Stiff soil with 600 ft/sec < V_s < 1200 ft/sec (183 m/sec < V_s < 365 m/sec)

(*d*) A soil profile with V_s < 600 ft/sec (183 m/sec) or any profile with more than 10 ft (3 m) of soft clay defined as soil with PI > 20, w > 40 percent, and $S_u \le 500$ psf (24 kpa).

V_s refers to shear wave velocity in all the soil types. The parameter A_a in Table 5.1 is the ground acceleration for rock with no soil and g is acceleration due to gravity.

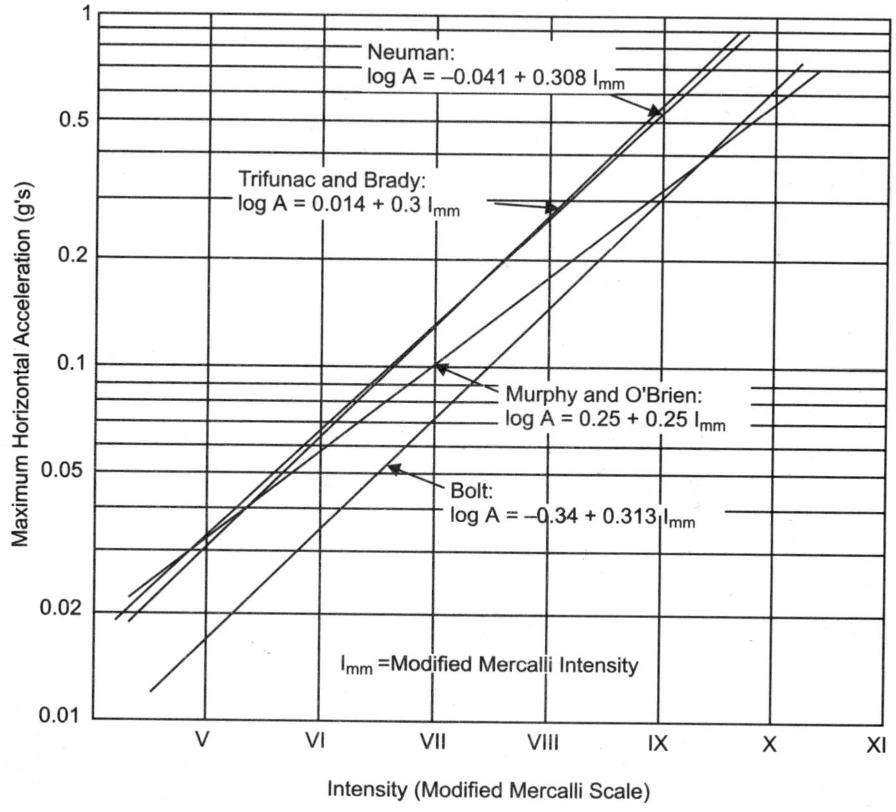

Fig. 5.5 approximate relationships between maximum acceleration and modified mercalli intensity (Courtesy: http://www.vulcanhammer.net)

Table 5.1 Peak ground acceleration modified for soil conditions (Courtesy: http://www.vulcanhammer.net)

Soil Type	*Aa = 0.1g*	*Aa = 0.2g*	*Aa = 0.3g*	*Aa = 0.4g*	*Aa = 0.5g*
A	0.1	0.2	0.3	0.4	0.5
B	0.12	0.24	0.33	0.4	0.5
C	0.16	0.28	0.36	0.44	0.5
D	0.25	0.34	0.36	0.46	SI

A_a = Effective peak acceleration

SI = Site specific geotechnical investigation and dynamic site response analyses are performed.

Note: Use straight line interpolation for intermediate values of A_a.

5.3.7 Magnitude and Intensity Relationships

For purposes of engineering analysis it may be necessary to convert the maximum intensity of the earthquake to magnitude. The most commonly used formula is:

$$M = 1 + \left(\frac{2}{3}\right) I_{MM} \quad \text{...(5.2)}$$

M represents magnitude and I_{MM} represents intensity. The above formula was derived to fit a limited database. The database primarily composed of Western United States earthquakes. It does not account for the difference in geologic structures or for depth of earthquakes, which may be important in the magnitude-intensity relationship.

5.3.8 Reduction of Foundation Vulnerability to Seismic Loads

In cases where potential for soil failure is not a factor, foundation ties, and special pile requirements can be incorporated into the design. The function of foundation ties and special pile requirements is to reduce the vulnerability to seismic loads. Details on these are given in NAVFAC P-355.1. In cases where there is a likelihood for soil failure (e.g., liquefaction), the engineer should consider employing soil improvement techniques.

Home Work Problems

1. With neat sketch explain about hypocenter, epicenter, focal depth and epicentral distance.
2. Explain about evaluation procedure for seismic soil response.
3. Discuss about design earthquake magnitude as well as the selection of magnitude level.
4. Write short notes on peak horizontal acceleration and magnitude-intensity interrelation.

6

CHAPTER

LIQUEFACTION

6.1 INTRODUCTION

The October 17, 1989 Loma Prieta earthquake was responsible for 62 deaths and 3,757 injuries. In addition, over $6 billion in damage was reported. This damage included damage to 18,306 houses and 2,575 businesses. Approximately 12,053 persons were reported displaced. The most intense damage was confined to areas where buildings and other structures where situated on top of loosely consolidated, water saturated soils. Loosely consolidated soils tend to amplify shaking and increase structural damage during earthquake. Water saturated soils compound the problem due to their susceptibility to liquefaction. Consequently, there is loss of bearing strength.

Liquefaction is a physical process that takes place during some earthquakes. Liquefaction may lead to ground failure. As a consequence of liquefaction, soft, young, water-saturated, well sorted, fine grain sands and silts behave as viscous fluids. This behaviour is very different than solids behaviour. Liquefaction takes place when seismic shear waves pass through a saturated granular soil layer. These shear waves distort its granular structure, and cause some of its pore spaces to collapse. The collapse of the granular structure increases pore space water pressure. Furthermore, it also decreases the soil's shear strength. If pore space water pressure increases to the point where the soil's shear strength can no longer support the weight of the overlying soil, buildings, roads, houses, etc., then the soil will flow like a liquid. Consequently, extensive surface damage results.

Fortunately, areas susceptible to liquefaction can be readily identified and the hazard can often be mitigated. Because of the relative ease of identifying hazardous areas, numerous liquefaction maps have been made by govenrnment agencies. Liquefiable sediments are young, loose, water saturated, and well sorted. They are either fine sands or silts. The sediments are seldom older than Holocene, and are usually only present on the modern floodplains of creeks and rivers. The map in Fig. 6.1 identifies areas likely to liquefy during a big earthquake (noted by the letter "L")

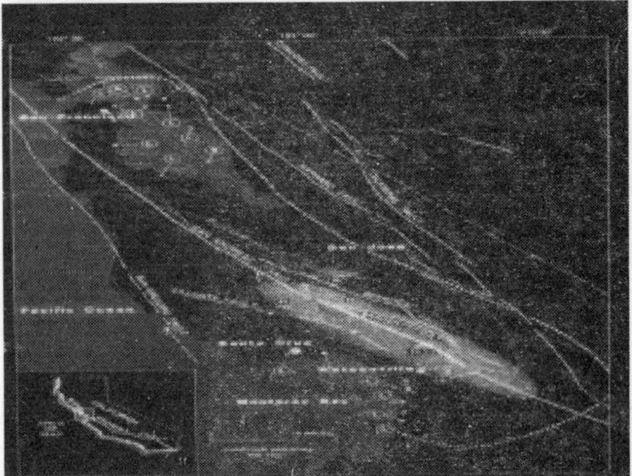

Fig. 6.1 Areas likely to liquefy during big earthquake (Courtesy: http://www.es.ucsc.edu)

In many low-lying coastal areas sand volcanoes are formed where underlying saturated sands liquefy during the seismic shaking. Furthermore, it ejects onto the surface. During the 1906 earthquake, sand volcanoes spouted to a heights around 20 feet in the Salinas Valley and near Moss Landing. During the Loma Prieta earthquake, extensive liquefaction occurred along the entire shoreline of the Monterey Bay, as well as in San Francisco's Marina District and along the bayshore in Oakland (Fig. 6.2).

Fig. 6.2 Sand Volcanoes (Courtesy: http://www.es.ucsc.edu)

Violent ground shaking combined with liquefaction of unconsolidated slough muds led to the spectacular failure of this bridge (Fig. 6.3). The bridge was on the famous Pacific Coast Highway 1 near Watsonville. The portions of the highway that collapsed were directly over saturated slough sediments. Upward acceleration of the bridge during the shaking caused the structure to separate from its support columns. The bridge then fell back downward, and the columns punctured through the concrete road surface.

Fig. 6.3 Failure of bridge due to liquefaction (Courtesy: http://www.es.ucsc.edu)

Collapsed highway bridge (close-up of support columns), State Highway 1, Struve Slough, Watsonville is shown in Fig. 6.4. Visible to the left of the columns in this figure are skid and scrape marks from a California Highway Patrol car. The car flew onto the bridge at high speed minutes after the earthquake. The driver, who survived the incident, was unaware that the bridge had collapsed as he raced down the highway in response to an earthquake emergency.

Fig. 6.4 Failure of bridge due to liquefaction, close-up view (Courtesy: http://www.es.ucsc.edu)

Collapsed highway bridge (close-up of support columns from below), State Highway 1, Struve Slough, Watsonville is shown in Fig. 6.5. This view from below the collapsed bridge shows the area around the base of a support column where soil was pushed away by the back and forth motion of the column during the earthquake. The rebar at the top of the column is still attached to the original location of contact between the column and the bridge. It has already been reported that the upward acceleration of the bridge during the shaking caused the structure to separate from the support columns. As the bridge then fell back downward, the columns punctured through the concrete road surface.

Liquefaction induced road failure, Moss Landing State Beach is shown in Fig. 6.6 (a). This road was built across an estuary and suffered extensive damage due to liquefaction. It subsided several meters during the earthquake. As a result, it got separated from the adjoining sections of road. The exposed cross section of underlying sediments in Fig. 6.6(b) clearly shows a lower light colored beach sand that liquefied and injected into an overlying dark silty unit.

Fig. 6.5 Failure of bridge due to liquefaction, close-up view from below (Courtesy: http://www.es.ucsc.edu)

Fig. 6.6(a) Liquefaction induced road failure (Courtesy: http://www.es.ucsc.edu)

Fig. 6.6(b) Liquefaction induced road failure, exposed cross section of underlying soils (Courtesy: http://www.es.ucsc.edu)

Failed river dike, San Lorenzo River, Beach Flats, Santa Cruz due to liquefaction is shown in Fig. 6.7. Much of the earthen San Lorenzo River flood control levee in the Beach Flats area of Santa Cruz were found to slumped and fractured as a result of extreme ground shaking, as well as due to liquefaction of underlying unconsolidated beach and river sediments.

Fig. 6.7 Failed river dike due to liquefaction (Courtesy: http://www.es.ucsc.edu)

Failure and cracks induced by liquefaction have also been observed. These images (Fig. 6.8(a) and Fig. 6.8(b)), probably from the 1906 earthquake event, shows cracks formed by liquefaction at the San Lorenzo River bed.

Fig. 6.8(a) Crack formation due to liquefaction (Courtesy: http://www.es.ucsc.edu)

Fig. 6.8(b) Crack formation due to liquefaction (Courtesy: http://www.es.ucsc.edu)

The concept of liquefaction was first introduced by Casagrande. Typical subsurface condition that is susceptible to liquefaction is newly deposited or placed loose sand. Furthermore, groundwater table should be near ground surface. During an earthquake, cyclic shear stresses are applied to loose sand. These cyclic shear stresses are induced due to propagation of shear waves during earthquake. Consequently, loose sand contracts. This results in increase in pore water pressure. Seismic shaking takes place too quickly. Consequently, cohesionless soil is subjected to undrained loading. This is total stress analysis condition. The increase in pore water pressure during earthquake thus causes upward flow of water to ground surface. At the ground surface it emerges as mud spouts or sand boils. Consequently, development of high pore water pressure due to ground shaking caused by earthquake and upward flow of water turns sand into liquefiable condition. This mechanism is called liquefaction. At the state of liquefaction, the effective stress is zero. Furthermore, individual soil particles are released from confinement. It appears that soil particles are floating in water.

Structures on top of loose sand deposit which has liquefied during earthquake will tend to sink. It also tends to fall over. Buried tanks will tend to float on the surface when loose sand liquefies. There are other important effects of liquefaction as well.

After the sand has liquefied, the excess pore water will start to dissipate. There are two important factors governing the duration for which soil will remain in liquefied state. First is the duration of seismic shaking during earthquake. Second is the drainage conditions of liquefied soil. The longer the cyclic shear stress application from the earthquake, longer the state of liquefaction persists. Similarly, stronger the cyclic shear stress application from earthquake, longer the state of liquefaction. Furthermore, if the liquefied soil is confined by an upper and lower clay layer, it will take longer duration for excess pore water pressure to dissipate. This is accomplished by flow of water from liquefied soil. After liquefaction process is complete, soil is in somewhat denser state.

Liquefaction can result in ground surface settlement. It can even result in bearing capacity failure of foundation. Liquefaction can also cause or contribute to lateral movement of slopes.

Liquefaction of soils has been extensively studied in laboratory. There is considerable amount of published data available concerning laboratory liquefaction testing. As per Ishihara (1985), laboratory tests were performed on hollow cylindrical specimens of saturated river sand. Testing was done in torsional shear test apparatus. Two samples were tested. First was having relative density 47% and the other was having relative density 75%. Prior to cyclic shear testing, both specimens were subjected to same effective confining pressure. Both samples were then subjected to undrained conditions during cyclic shear stress application. Cyclic shear stress had constant amplitude and sinusoidal pattern in both samples. Sand having lower relative density was subjected to lower constant amplitude cyclic stress than the sample having higher relative density (amplitude ratio of lower to higher relative density specimen = 0.32). For sand having lower relative density, there was sudden and rapid increase in shear strain (as high as 20%) during constant amplitude cyclic shear stress application. On the other hand in sand having higher relative density, shear strain was found to slowly increase with application of constant amplitude cyclic shear stress. During application of cyclic shear stress, excess pore water pressure was found to develop in both the specimens. When the excess pore water pressure becomes equal to initial effective confining pressure, the effective stress becomes zero. Consequently, shear strain dramatically increases. This is liquefaction condition. The condition of effective stress zero and sudden shear strain increase (i.e. liquefaction condition) was observed in sand having low relative density. For sand having high relative density, effective stress became zero due to cyclic shear stress application. However, sudden shear strain increase was not observed. This happens because on reversal of cyclic shear stress, dense sand dilates. Consequently, undrained shear resistance increases. Liquefaction is thus a momentary condition in dense sands. This mechanism in dense sands is also called cyclic mobility. Furthermore, in sand having lower relative density, there was permanent loss in shear strength with each additional cycle of constant amplitude shear stress. No such observation was made in sand having higher relative density. Earthquakes will not subject soil to uniform constant amplitude cyclic shear stress. However, it can be modeled accordingly. Consequently, experimental results as explained above are of great importance.

As per See and Lee 1965, cyclic triaxial tests were performed. Testing was done on saturated specimens of river sand. Cylindrical specimens were first saturated. Then they were subjected in triaxial apparatus to an isotropic effective confining stress. This stress was of constant magnitude. Samples were then subjected to undrained condition during cyclic deviator stress application. Numerous samples were tested. Testing was done at different void ratios as well as at different cyclic deviator stress amplitude value. Number of cycles required for initial liquefaction and 20% axial strain was determined under each condition. In samples having same initial void ratio and same initial effective confining stress, with increase in deviater cyclic stress amplitude, there was decrease in number of cycles required for liquefaction. For a sample having same initial void ratio and same effective confining pressure, cyclic deviator stress amplitude required to cause liquefaction decreases as the number of cycles of deviator stress is increased. A dense sand was found to have greater resistance for liquefaction. It would require higher deviator cyclic stress amplitude or more number of cycles of deviator cyclic stress to cause initial liquefaction as compared to same soil in loose state (Seed and Lee, 1966).

Finally, based on above two experimental observations, it can be concluded that liquefaction is more dominant in loose sands.

6.2 FACTORS GOVERNING LIQUEFACTION IN THE FIELD

There are many factors governing liquefaction of in-situ soil.

1. In order to have liquefaction, there must be ground shaking. Character of ground motion, such as acceleration and duration of shaking, determines shear strains. Shear strains cause contraction of soil particles. Furthermore, it also leads to development of excess pore pressure. These activities lead to liquefaction. Most common cause of liquefaction is seismic energy released during earthquake. Potential for liquefaction increases with increase in earthquake shaking. It also increases due to duration of shaking. It has been reported that shaking threshold needed for liquefaction is peak ground acceleration of 0.1g and local earthquake magnitude of about 5. Consequently, liquefaction analysis is not needed at sites having peak ground acceleration less than 0.1g and local earthquake magnitude less than 5 during earthquakes.
2. The condition most conducive to liquefaction is near surface ground water table location. Unsaturated soil located above groundwater table will not liquefy. If at a particular site, soil is currently above groundwater table and is highly unlikely to get saturated for given foreseeable changes in hydrologic regime, such soil need not be evaluated for liquefaction potential. If groundwater table at a particular site fluctuates significantly, liquefaction potential also fluctuates. In general, highest historic groundwater level should be used in liquefaction analysis. Liquefaction can also occur in very large masses of sands or silts that are dry and loose. If such soils are loaded so rapidly that escape of air from voids is restricted, they can liquefy. This mechanism

of liquefaction in dry and loose sand is called running soil or running ground. However, contribution to liquefaction by this mechanism is not significant. Technically, the term liquefaction is used only for soils located below groundwater table level.

3. Soil types most susceptible to liquefaction during earthquake is deposit consisting of fine to medium sand. Sands containing low plasticity fines are also susceptible to liquefaction. Some times liquefaction takes place in gravelly soils as well (Kuwabara and Yoshumi, 1973). Consequently, all the soils susceptible to liquefaction are cohesionless soils. Approximate listing of cohesionless soils from least to most resistant to liquefaction is clean sands, nonplastic silty sands, nonplastic silt and gravels. However, there are exceptions. Tailings derived from mining industry are essentially composed of ground-up rocks. They are classified as rock flour. Rock flour in water saturated state is not having significant cohesion. It behaves like clean sand. Similar to clean sand they are found to have low resistance to liquefaction. Based on laboratory experiments, it has been observed that during earthquakes great majority of cohesive soils in general do not liquefy. This fact has been verified by field observations as well. For cohesive soil to liquefy it must have 15% of particles (based on dry weight) less than 0.005 mm particle size, liquid limit less than 35% and water content greater than 0.9 times the liquid limit. If any one of the criteria is not met, cohesive soil is not considered to be susceptible to liquefaction. Although cohesive soil may not liquefy, there could be significant undrained shear strength loss due to seismic shaking during earthquakes.

4. Cohesionless soils in loose relative density state are susceptible to liquefaction. Loose nonplastic soils will contract during seismic shaking. This causes development of excess pore water pressure. Upon reaching liquefaction, there will be sudden and dramatic increase in shear displacement for these soils. For dense sands, initial liquefaction does not produce large deformations. This is due to dilation tendency of sand upon reversal of cyclic shear stress. Consequently, dilative soils need not be evaluated for liquefaction since they are not susceptible to liquefaction. Dilative soils are not susceptible to liquefaction because their undrained shear strength is more than their drained shear strength.

5. Uniformly graded nonplastic soils form more unstable particle arrangements. They are more susceptible to liquefaction than well graded soils. Well-graded soils have particles filling the void spaces between large particles. This reduces potential contraction of soil, resulting in less excess pore water pressure being generated during earthquake. Even field observations indicate that most liquefaction failure involve uniformly graded granular soils.

6. Hydraulic fill (fill placed under water), tend to be more susceptible to liquefaction because of loose and segregated soil structure. This kind of structure is created by soil particles falling through water. Natural soil deposits formed in lakes, rivers or oceans are also susceptible to liquefaction due to same reason. Soils which are especially susceptible to liquefaction are formed in lacustrine, alluvial and marine depositional environments.

7. If excess pore pressure can quickly dissipate, soil may not liquefy. Consequently, highly permeable gravel drains or gravel layers can reduce liquefaction potential of adjacent soil as they will help in dissipating pore pressure rapidly.
8. Greater the confining pressure, lesser the soil has tendency to liquefy. Deeper groundwater table location, soil located at deeper depth below ground surface and surcharge pressure applied at ground surface are the conditions creating a higher confining pressure. It has been reported that possible zone of liquefaction extends from ground surface to a maximum depth of 15 m. Deeper soils usually don't liquefy due to higher confining pressure. However, that doesn't mean that liquefaction analysis should not be performed for a soil located below a depth of 15 m. For example in sloping ground, such as sloping front in the form of water front structure should be analyzed for liquefaction. Similarly sloping shell of earthen dam are places where soil layer below 15 m depth should be analyzed for liquefaction. Liquefaction analysis should be performed for the entire thickness of soil deposit that has been loosely dumped in water even if it exceeds 15 m thickness. Finally it can be said that experience and judgement is required in the determination of proper depth at which liquefaction analysis should be terminated.
9. Soils having rounded particles tend to densify more easily than angular shaped soil particles. Consequently, rounded soil particles is more susceptible to liquefaction than soil containing angular soil particles during earthquake.
10. New soil deposits are more susceptible to liquefaction than old deposits. Longer a soil is subjected to confining pressure, greater the liquefaction resistance. Consequently, older deposits have greater liquefaction resistance. Increase in liquefaction resistance with time is due to deformation or compression of soil particles into more stable arrangements. There may also be development of bonds due to cementation at particle bonds with time.
11. Historical environment of soil can affect its liquefaction potential. Older soil deposits that have already been subjected to seismic shaking have increased liquefaction resistance. However, newer deposits of same soil at same density not subjected to seismic shaking are more prone to liquefaction. Liquefaction resistance also increases, with increase in overconsolidation ratio. Furthermore, it also increases with increase in coefficient of earth pressure at rest. Consequently a soil that has been preloaded, has a higher resistance to liquefaction than the soil which has not been preloaded. Underlying soil when the upper layer has been removed by erosion is one such example which has been preloaded.
12. Construction of heavy building on top of sand deposit decreases liquefaction resistance of soil. Soil is subjected to shear stress caused by building load. Soil requires smaller additional load from the earthquake to cause contraction. Contraction causes liquefaction. Consequently, liquefaction resistance of soil decreases in the presence of building load. Building load also must be considered in liquefaction induced settlement. Building load also must be considered in liquefaction induced bearing capacity as well as in stability analysis.

6.3 LIQUEFACTION ANALYSIS

The first step in liquefaction analysis is to determine whether soil has ability to liquefy during earthquake. Most of the soils that are susceptible to liquefaction are cohesionless. Cohesive soils liquefy only under specific conditions. These conditions have been discussed in previous section. Most common type of analysis to determine liquefaction potential is the use of standard penetration test (Seed and Idriss, 1971). There is another widely used technique of liquefaction analysis. This is called simplified procedure. The steps involved in simplified procedure are as follows:

1. First step is to determine if the soil has ability to liquefy.
2. The soil must be below groundwater table. Liquefaction analysis can also be performed if it is anticipated that groundwater table will rise in future so that eventually soil will be below groundwater table.
3. Next step of simplified procedure is to determine the cyclic stress ratio (CSR). Cyclic stress ratio is induced during earthquake. Determination of cyclic stress ratio requires information about peak horizontal ground acceleration. Peak horizontal ground acceleration is caused due to earthquake.
4. By using the standard penetration test, the cyclic resistance ratio (CRR) of the soil is determined. If the CSR induced by earthquake is greater than CRR determined from standard penetration test, liquefaction of soil will take place during earthquake and vice versa.
5. Finally factor of safety against liquefaction is determined as FS = CRR/CSR. FS represents factor of safety.

6.3.1 Cyclic Stress Ratio (CSR) Caused by Earthquake

To determine CSR earthquake equation, it is assumed that ground surface is level. Furthermore, soil column is assumed to have unit width. Length of soil column is assumed to move horizontally as a rigid body due to maximum horizontal acceleration exerted by earthquake at ground surface (refer Fig. 6.9).

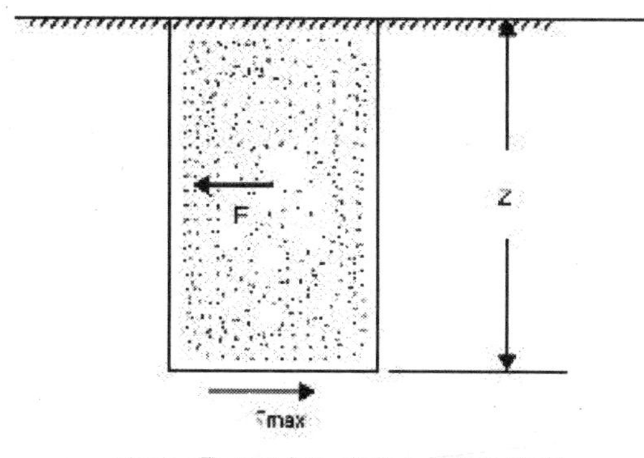

Fig. 6.9 Conditions assumed for the derivation of CSR earthquake equation (Courtesy: Day, 2002)

As per conditions of Fig. 6.9, horizontal earthquake force acting on soil column of unit width and length is:

$$F = ma = \left(\frac{W}{g}\right)a = \left(\frac{\gamma_t z}{g}\right)a_{max} = \sigma_{v0}\left(\frac{a_{max}}{g}\right) \quad ...(6.1)$$

where, F = horizontal earthquake force acting on soil column of unit width and length.

m = total mass of soil column.

W = total weight of soil column.

γ_t = total unit weight of soil.

z = depth below ground surface of soil column. (Fig. 6.9)

a = acceleration = maximum horizontal acceleration at surface due to earthquake (= a_{max}).

a_{max} = maximum horizontal acceleration at ground surface induced by earthquake.

σ_{v0} = total vertical stress at bottom of soil column = $\gamma_t z$.

This force is acting on soil element of unit width and length. Furthermore, it is equal to maximum shear stress (τ_{max}) at the base of soil element. Hence:

$$\tau_{max} = \sigma_{vo}\left(\frac{a_{max}}{g}\right) \quad ...(6.2)$$

Dividing both sides by σ'_{v0}, vertical effective stress, following is obtained:

$$\frac{\tau_{max}}{\sigma'_{v0}} = \left(\frac{\sigma_{v0}}{\sigma'_{v0}}\right)\left(\frac{a_{max}}{g}\right) \quad ...(6.3)$$

However, in reality soil column is not acting as rigid body. Furthermore, soil is deformable. Consequently, equation (6.3) is multiplied with depth reduction factor (r_d) (Yoshimi, 1967). In the simplified method, typical irregular earthquake record is converted to equivalent series of uniform stress cycles. This is done by assuming $\tau_{cyc} = 0.65\ \tau_{max}$. τ_{cyc} is uniform cyclic shear stress amplitude of earthquake. Cyclic stress ratio (CSR) is obtained as follows:

$$CSR = \left(\frac{\tau_{cyc}}{\sigma'_{v0}}\right) = 0.65 r_d \left(\frac{\sigma_{v0}}{\sigma'_{v0}}\right)\left(\frac{a_{max}}{g}\right) \quad ...(6.4)$$

Depth reduction factor varies with depth below ground surface. It is obtained from standard charts as a function of depth for a particular magnitude earthquake. Sometimes linear relation between depth reduction factor and depth is also assumed. This relation is given as, $r_d = 1 - 0.012z$. z is depth in meters and is measured from ground surface. σ_{v0} and σ_{v0} in equation (6.4) can be obtained from basic geotechnical analysis. Peak ground acceleration (a_{max}) caused by earthquake has already been discussed. It is obtained from seismogram data. Consequently, equation (6.4) can be used for CSR determination for a particular magnitude earthquake.

6.3.2 Cyclic resistance ratio (CRR) from standard penetration test

Cyclic resistance ratio represents liquefaction resistance of in situ soil. Standard penetration test data is used for this purpose. There are several advantages of using standard penetration test to evaluate liquefaction potential. Boring is excavated to perform the standard penetration test. Groundwater table location can be observed and measured in the bore hole. In clean sand, while doing standard penetration testing, sampler will not be able to retain soil sample. For other soils, soil can be retained in the sampler. Furthermore, it can be visually classified to estimate percentage of fines. Then it can be brought in laboratory for further classification study to assess liquefaction susceptibility of soil. Furthermore, factors increasing liquefaction resistance also increase the standard penetration value. Well graded dense sands are more resistant to liquefaction and have high standard penetration value. On the other hand uniformly graded soil with loose or segregated soil structure is more susceptible to liquefaction. Consequently, these soils have lower standard penetration value. It has been reported that for standard penetration value between 0 and 20, soil is highly susceptible to liquefaction. For standard penetration value larger than 30, there is no significant damage due to liquefaction. Fig. 6.10 can be used to determine cyclic resistance ratio (CRR). For most of the data of Fig. 6.10, the earthquake magnitude is close to 7.5 (Seed et al, 1975). Three lines of the figure are for soils containing 35, 15 or 5 percentage fines. The line is at boundary. Data to the left of line indicates liquefaction during earthquake. Data to the right indicates no liquefaction during earthquake.

$(N_1)_{60}$ in the x-axis of Fig. 6.10 represents field standard penetration value. This value has been corrected for field testing as well as for overburden. In order to use Fig. 6.10, this parameter is determined first from standard penetration testing. Estimated percentage of fines in the sample is then determined by performing field and laboratory analysis. Soil with more number of fines have higher liquefaction resistance. Fig. 6.10 is applicable for nonplastic silty sands and plastic silty sands. It is also applicable to cohesive soils meeting the criteria of liquefaction. By knowing $(N_1)_{60}$ and estimated percentage of fines, CRR of the soil can be estimated from Fig. 6.10. From Fig. 6.10 it is clear that for $(N_1)_{60}$ value greater than 30, clean sand will not liquefy for an earthquake of magnitude about 7.5. This value of $(N_1)_{60}$ corresponds to sand in dense or very dense state. Fig. 6.10 is applicable for earthquake magnitude of about 7.5. As the earthquake magnitude increases, duration of ground shaking also increases. Consequently, cyclic shear strain will increase. This decreases liquefaction resistance of soil. For earthquake magnitudes other than 7.5, the CRR obtained from Fig. 6.10 is multiplied by magnitude scaling factor. Magnitude scaling factor has been given in Table 6.1.

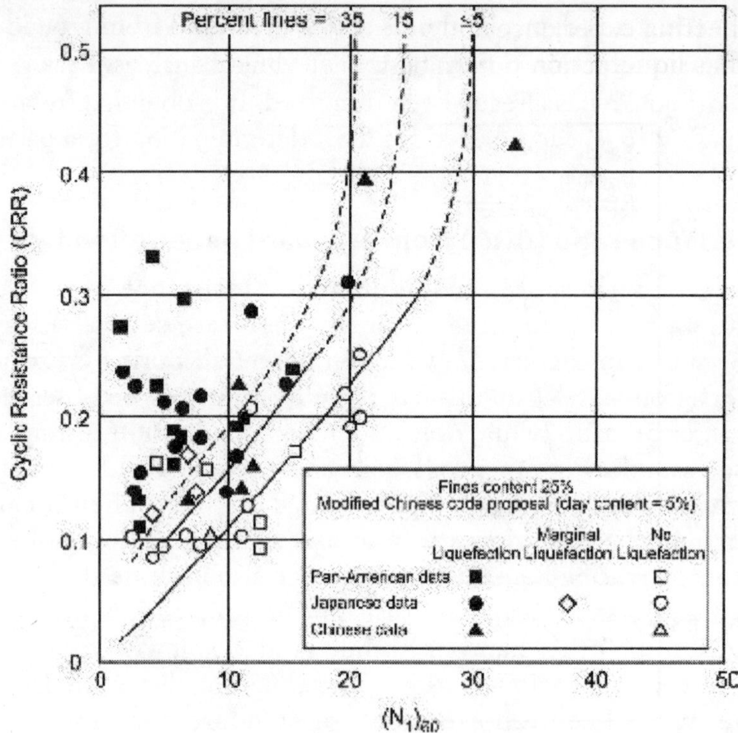

Fig. 6.10 Plot used to determine cyclic resistance ratio for clean and silty sands for M = 7.5 earthquake (Courtesy: Day, 2002)

Table 6.1 Magnitude scaling factors (Courtesy: Day, 2002)

Anticipated earthquake magnitude	*Magnitude scaling factor*
8.5	0.89
7.5	1.00
6.75	1.13
6	1.32
5.25	1.50

Factor of safety against Liquefaction: Factor of safety (FS) against liquefaction is defined as: FS = CRR/CSR. If the cyclic stress ratio caused by anticipated earthquake is greater than cyclic resistance ratio of in-situ soil, liquefaction could occur during earthquake. Liquefaction will not take place otherwise. Higher the factor of safety, more is the resistance of soil against liquefaction during earthquake. Soil having factor of safety slightly greater than one can also liquefy. For example if lower layer liquefies, then upward water flow could induce liquefaction of upper layer as well. This layer has factor of safety against liquefaction slightly greater than one.

However, in the above analysis, there are lot of corrections. These corrections are applied both to cyclic stress ratio as well as to cyclic resistance ratio. This is done for more accurate analysis. Otherwise the entire analysis is only gross approximation. Consequently, considerable engineering experience and judgement are essential in the final determination of whether a site has liquefaction potential or not.

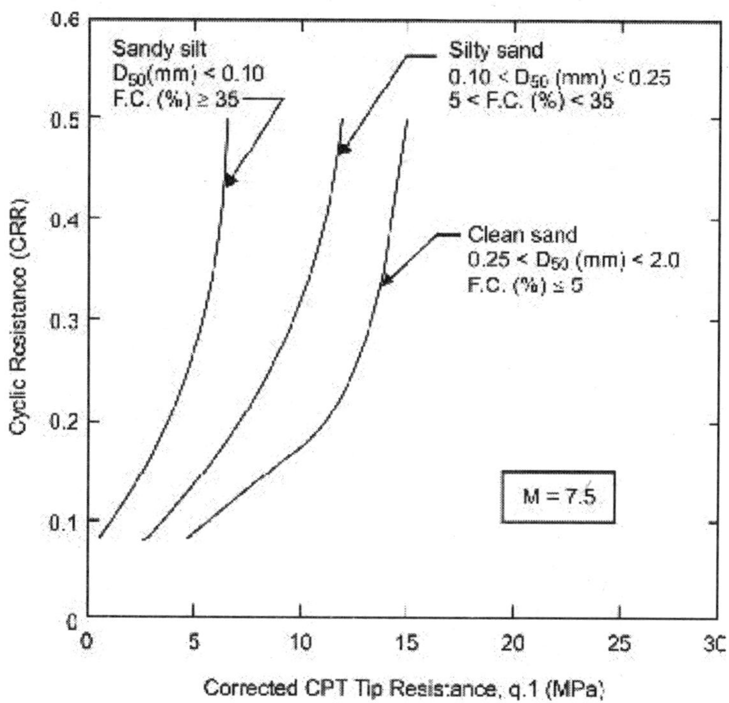

Fig. 6.11 Relation between CRR and corrected CPT tip resistance value for clean sand, silty sand and sandy silt for M = 7.5 earthquake (Courtesy: Day, 2002)

There are alternate techniques of determining CRR as well. In one technique corrected cone penetration test (CPT) tip resistance is determined. Then cyclic resistance is determined as its function by using Fig. 6.11. Fig. 6.11 is applicable for earthquake magnitude about 7.5. CRR value obtained from Fig. 6.11 should be multiplied by magnitude scaling factor for other magnitude earthquakes. Magnitude scaling factor is available in Table 6.1. Fig. 6.11 also has different curves that are to be used depending on percentage of fines in soil. It is clear from Fig. 6.11 that soils with more fines have higher cyclic resistance ratio for a given corrected CPT tip resistance value.

In the shear wave velocity method, shear wave velocity is determined by using standard geophysical techniques. Uphole, downhole, cross hole, seismic cone penetrometer and suspension logger etc are example of these geophysical techniques. It is an in-situ technique. In-situ shear wave velocity thus determined is corrected for overburden using $V_{s1} = V_s\left(\frac{100}{\sigma'_{v0}}\right)^{0.25}$. V_{s1} is corrected shear wave velocity. V_s is in-situ shear wave velocity. σ'_{v0} is vertical effective stress (kPa). Fig. 6.12 is then used to determine cyclic resistance ratio. Advantage of using shear wave velocity to determine factor of safety against liquefaction is that it can be used for large sites. Furthermore, it is also useful at places where initial assessment of liquefaction potential is required.

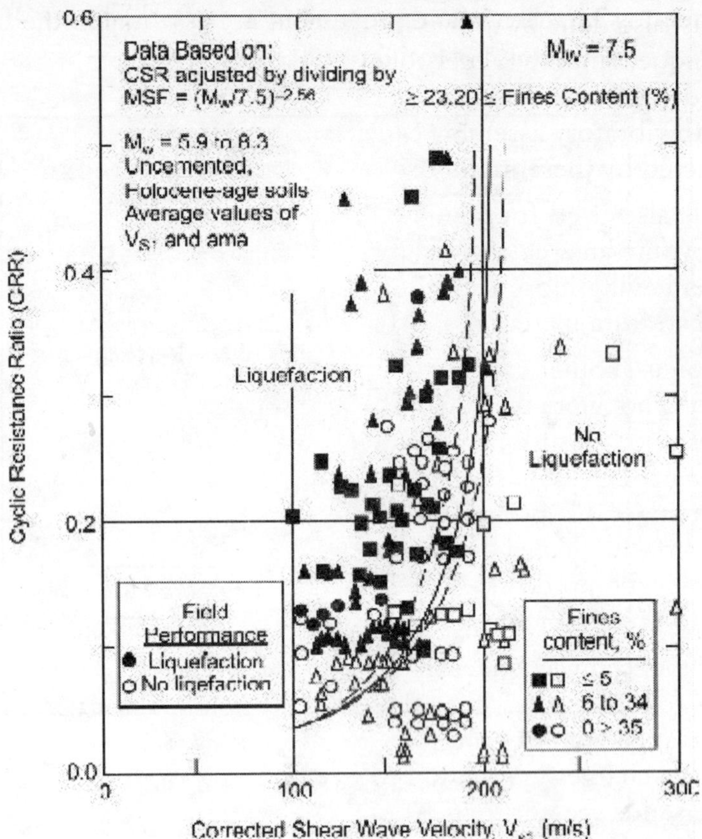

Fig. 6.12 Relation between cyclic resistance ratio and corrected shear wave velocity for clean sand, silty sand and sandy silt for M = 7.5 earthquake (Courtesy: Day, 2002)

6.4 ANTILIQUEFACTION MEASURES

A comprehensive study is required to find out various measures to prevent liquefaction. It depends on lot of factors. A few of them can be controlled in field. Based on these studies certain methods have been suggested (Lew, 1984). Liquefaction resistance can be improved to some extend by using methods given below.

6.4.1 Compaction of Loose Sands

Loose saturated sands are more prone to liquefaction. However, dense saturated sands have more resistance for liquefaction. Consequently, the liquefaction potential can be reduced by compacting the loose sand deposit. This should be done before any structure is constructed. There are different methods of compacting loose sand.

One technique is rolling loose sand with rubber tyre rollers. This is accomplished by excavating loose sand to some depth. Backfilling is then done in controlled lift thickness. Afterwards compaction of the loose sand is done. While using rubber tyres, lifts are usually 150 mm to 200 mm. However, this method cannot be used for compacting deep sand deposits.

Compaction is also done by using vibratory plates. Sometimes vibratory rollers are also used. In this technique, smooth wheel rollers are used. They are provided with vibratory device inside. Lift depths upto about 1.5 m to 2 m can be compacted with this equipment. Plates mounted with vibratory assembly can also be used. However, only small thickness of soils can be compacted by these methods. Technique is not useful for large deposits.

Pile driving is also used for compaction. Piles when driven in loose sand deposits, compacts the sand within an area covered by eight times around it. This technique is utilized in compacting sites having loose sand deposits. Since pile remains in sand, the over all stiffness of the soil stratum increases substantially due to pile driving.

Vibrofloatation is another compaction technique. It is used in cohesionless deposits of sand and gravel having not more than 20% silt or 10% clay. Vibrofloatation utilizes a cylindrical penetrator which is about 4 m long and 400 mm in diameter. The lower half is vibrator. Upper half is stationary. Device has water jets at top and bottom. Vibrofloat is lowered under its own weight. Bottom jet is kept on. This induces quick sand condition. When the vibrofloat reaches desired depth, the flow is diverted to upper jet and vibrofloat is pulled out slowly. Top jet aids the compaction process. As the vibrofloat is pulled out, a crator is formed. Sand or gravel is added to the crator formed.

Blasting is another compaction technique. Explosion of buried charge induces liquefaction of soil mass. This is followed by escape of excess pore water pressure. This acts as lubricant and facilitates re-arrangement of sand particles. This leads sand to more compacted state. Lateral distribution of charges in ground is based on results obtained from a series of single shots. Where loose sands greater than 10 m thick are to be compacted, two or more tyres of small charges are preferred. For deposits less than 10 m thick, charges placed at 2/3rd depth from surface is generally sufficient. There is no apparent limit of depth that can be compacted by means of explosive (Lyman, 1942). Repeated blasts are found to be more effective than a single blast of several small charges. These charges are detonated simultaneously. Very little compaction is achieved in top 1 m due to blasting by large charge. Small charges are found to be more effective than large charges for compacting upper 1.5 m sand. Compaction gained by repeating the blasts more than 3 times is found to be small. Relative density can be increased to 80% by blasting.

6.4.2 Grouting and Chemical Stabilization

In grouting, some kind of stabilizing agent is inserted into the soil mass. This is done under pressure. The pressure forces the stabilizing agent into soil voids in a limited space. This limited space is around the injection tube. The stabilizing agent either reacts with soil or with itself to form a stable soil mass. The most common type of stabilizing agent (also called grouting agent) is a mixture of cement and water. It may or may not contain sand. Generally grout can be used if the permeability of the deposit is greater than 10^{-5} m/s. In chemical stabilization, lime, cement, flyash or their combination is used as stabilizing agent.

6.4.3 Application of Surcharge

Application of surcharge over the deposit liable to liquefy can also be used as an effective measure against liquefaction. Fig. 6.13 shows a plot between rise in pore water pressure and effective overburden pressure at an acceleration of 10 percent of acceleration due to gravity. From the figure it can be seen that pore pressure increases with increase in overburden pressure till a maximum value of pore pressure is reached. Beyond this value of overburden pressure, further application of overburden pressure decreases the pore pressure value. Consequently, overburden pressure higher than the value corresponding to maximum pore pressure will make the deposit safe against liquefaction.

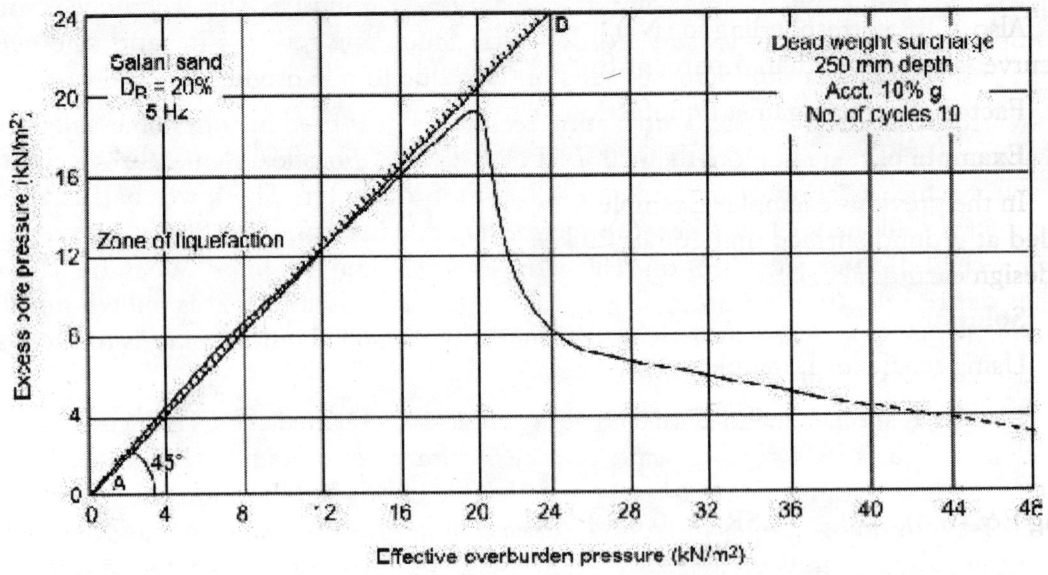

Fig. 6.13 Excess pore water pressure versus effective overburden pressure on Solani sand (Courtesy, Swami Saran, 1999)

6.4.4 Drainage Using Coarse Material Blanket and Drains

Blankets and drains of material with higher permeability reduce the length of drainage path. Furthermore, due to higher coefficient of permeability it also speeds up the drainage process. These activities help to make soil deposit safe against liquefaction (Katsumi et al, 1988 and Susumu et al, 1988).

Example 6.1

It is planned to construct a building on a cohesionless sand deposit (fines < 5 percent). There is a nearby major active fault, and the engineering geologist has determined that for the anticipated earthquake, the peak ground acceleration a_{max} will be equal to 0.45 g. At this location, ground surface is level, water table is 1.5 m below ground surface, total unit weight of soil above water table is 18.9 kN/m^3 and submerged unit weight of soil below water table is 9.84 kN/m^3. $(N_1)_{60}$ corresponding to 3m depth is 7.7. Assuming an anticipated earthquake magnitude of 7.5, calculate the factor of safety against liquefaction for the saturated clean sand at a depth of 3m below ground surface.

Solution:

At 3 m depth, at the given location,

$$\sigma_{v0} = (1.5)(18.9) + (1.5)(9.84+9.81) = 58 \text{ kPa}.$$

$$\sigma'_{v0} = (1.5)(18.9) + (1.5)(9.84) = 43 \text{ kPa.}$$

Using, $r_d = 1 - 0.012z$, with $z = 3m$, $r_d = 0.96$

Also, $$\frac{\sigma_{v0}}{\sigma'_{v0}} = 58/43 = 1.35 \text{ and } \frac{a_{max}}{g} = 0.45g/g = 0.45$$

Substituting the values in Eq. (6.4), CSR = 0.38

Also, CRR corresponding to $(N_1)_{60} = 7.7$ for 3m depth, using Fig. 6.10 and intersecting the curve labeled less than 5 percent fines, CRR = 0.09.

Factor of safety against liquefaction = CRR/CSR = 0.09/0.38 = 0.237

Example 6.2

In the previous example (Example 6.1), assume that there is vertical surcharge pressure applied at ground surface that equals 20 kPa. Determine the cyclic stress ratio induced by the design earthquake.

Solution:

Using results of Example 6.1,

$$\sigma_{v0} = 58 + 20 = 78 \text{ kPa.}$$

$$\sigma'_{v0} = 43 + 20 = 63 \text{ kPa.}$$

Using Eq. (6.4), $$CSR = 0.65(0.96)(78/63)(0.45) = 0.347$$

Home Work Problems

1. Solve Example 6.1 assuming $a_{max}/g = 0.1$ and the sand contains 15 percent non plastic fines. (**Ans.** Factor of safety = 1.67).
2. Solve Example 6.1 assuming $a_{max}/g = 0.3$ and the earthquake magnitude M = 5.25. (**Ans.** Factor of safety = 0.534).
3. Using data of Example 6.1, determine factor of safety against liquefaction assuming shear wave velocity = 150 m/s and $a_{max}/g = 0.1$. (Ans. Factor of safety = 1.9).
4. What is liquefaction? Explain with examples.
5. What are the factors governing liquefaction in field?
6. Develop cyclic stress ratio equation.
7. How is cyclic resistance ratio determined from shear wave velocity method?

EARTHQUAKE RESISTANT DESIGN OF SHALLOW FOUNDATION

7.1 INTRODUCTION

A bearing capacity failure is a foundation failure. This foundation failure occurs when the shear stresses in the soil exceed the shear strength of soil. For both static and seismic cases, bearing capacity failure is grouped into three categories (Vesic, 1973). They are called general, local and punching shear failure. In general shear failure, there is complete rupture of underlying soil. Furthermore, soil is pushed up on both sides. There is complete shear failure of soil. General shear failure takes place in soils which are in dense or in hard state. In punching shear failure, there is compression of soil directly below the footing. There is vertical shearing as well. Furthermore, soil outside the loaded area remains uninvolved. However, there is minimum movement of soil on both sides of footing. It occurs in soils that are in loose or soft state. Local shear failure can be considered as a transition phase between general shear and punching shear. There is rupture of soil only immediately below footing in this type of shear failure. There is small soil bulging on both sides of footing. Local shear failure takes place in soils which are in medium or firm state.

It has been reported that compared to damage by earthquake-induced settlement, there are fewer damage by earthquake-induced bearing capacity failure. There are several reasons for it. In most cases, settlement is found to be governing factor. Consequently, foundation bearing pressures recommended are based on limiting the amount of expected settlement. This recommendation is applicable to static as well as seismic conditions. There have been extensive studies of both static and seismic bearing capacity failure of shallow foundations. This has lead to development of bearing capacity equations. It has been suggested that for the evaluation of bearing capacity for seismic analysis, the factor of safety should often be in the range of 5 to 10. Larger footing size lowers the bearing pressure on soil. It also reduces potential for static or seismic bearing capacity failure.

Usually, there are three factors causing failure during earthquake. Overestimation of shear strength, as well as loss of shear strength due to liquefaction during earthquake is one factor. Secondly, earthquake causes rocking of the structure. Resulting structural overturning moments produce significant cyclic vertical thrusts on foundation. This increases the structural load. Thirdly, altered site due to earthquake can also produce bearing capacity failure. The most common cause of a seismic bearing capacity failure is liquefaction of underlying soil. For static analysis, soil involved in bearing capacity failure extends to a depth equal to footing width. However, this depth of bearing capacity failure might exceed for earthquake induced loading.

It is recommended that the allowable bearing pressure be increased by a factor of one-third under earthquake conditions. This recommendation is for seismic analysis under massive crystalline bedrock, sedimentary rock, dense granular soil or heavily overconsolidated cohesive soil conditions. However, this increase is not recommended for foliated rock, loose soil under liquefaction, sensitive clays and soft clays. Since, in these cases there is weaking of soil and hence allowable bearing pressure has to be reduced during earthquake.

7.2 BEARING CAPACITY ANALYSIS FOR LIQUEFIED SOIL

Table 7.1 summarizes the requirements and analyses for soil susceptible to liquefaction.

Table 7.1 Requirements and analyses for soil susceptible to liquefaction (Courtesy: Day, 2002)

Requirements and analyses	*Design conditions*
Requirements	1. Bearing location of foundation: The foundation must not bear on soil that will liquefy during earthquake.
	2. Surface layer: There must be adequate thickness of unliquefiable soil layer to prevent damage due to sand boils and surface fissuring.
Settlement analysis	1. Lightweight structures: Settlement of lightweight structures (wood-frame building on shallow foundation)
	2. Low net bearing stress: Settlement of any other kind of structure imparting low net bearing pressure.
	3. Floating foundation: Settlement of floating foundation below bottom of foundation provided zone of liquefaction is below foundation base and there is no net stress.
	4. Heavy structure with deep liquefaction: Settlement of heavy structures provided zone of liquefaction is deep enough that stress increase caused by structural load is low.
	5. Differential settlement: Differential settlement if structure contains deep foundation supported by strata below zone of liquefaction.
Bearing capacity analysis	1. Heavy building with underlying liquefied soil: Use adequate bearing capacity analysis assuming soil is liquefied due to earthquake. Foundation load will cause it to punch or sink in liquefied soil.

	2. Check bearing capacity: Perform bearing capacity analysis whenever footing imposes net pressure into soil and underlying soil layer is susceptible to liquefaction during earthquake. 3. Positive induced pore pressures: Perform bearing capacity analysis when soil will not liquefy during earthquake but there is development of excess pore pressure.
Special considerations	1. Buoyancy effects: To be considered for buried storage tank, large pipelines which may float on surface when soil liquefies. 2. Sloping ground condition: Determine if the site is susceptible to liquefaction induced flow slide.

In punching shear analysis, during earthquake loading it is assumed that load causes foundation to punch straight downward through upper unliquefiable soil layer down into liquefied soil layer. Factor of safety is considered as follows:

For strip footing:

$$FS = \frac{2T\tau_f}{P} \quad ...(7.1)$$

For spread footing:

$$FS = \frac{2(B+L)(T\tau_f)}{P} \quad ...(7.2)$$

where,

T = vertical distance from bottom of footing to top of liquefied soil layer.

τ_f = shear strength of unliquefiable soil layer.

B = width of footing.

L = length of footing.

P = footing load (dead, live, seismic loads and self weight of footing)

FS = factor of safety.

Shear strength of unliquefiable soil layer is determined using conventional techniques. This technique is applicable for cohesive as well as for cohesionless soils.

For cohesive Soil:

$$\tau_f = s_u \quad ...(7.3)$$

or,

$$\tau_f = c + \sigma_h \tan\phi \quad ...(7.4)$$

For cohesionless soil:

$$\tau_f = k_0\, \sigma'_{vo} \tan\phi' \quad ...(7.5)$$

where,

s_u = undrained shear strength of cohesive soil.

c, ϕ = undrained shear strength parameters.

σ_h = horizontal total stress.

k_0 = coefficient of earth pressure at rest.

σ'_{v0} = vertical effective stress $\left(\text{at } \frac{T}{2} + \text{footing depth from ground surface}\right)$.

σ' = effective friction angle of cohesionless soil.

For local and general shear failure conditions, Terzaghi bearing capcity equation is used. Furthermore, the basic equation is modified for different type of footing and loading conditions (Terzaghi, 1943 and Meyerhof, 1951)). For the situation of cohesive soil layer overlying sand which is susceptible to liquefaction, a total stress analysis is performed. Following equations are used:

For strip footing:

$$q_{ult} = s_u N_c \quad ...(7.6)$$

For spread footing:

$$q_{ult} = s_u N_c \left(1 + 0.3\frac{B}{L}\right) \quad ...(7.7)$$

where,

s_u = undrained shear strength.

N_c = bearing capacity factor determined from Fig. 7.1 for the condition of a unliquefiable cohesive soil layer overlying a soil layer that is expected to liquefy during design earthquake.

B = footing width.

L = footing length.

For liquefied soil layer, the shear strength value is zero (c_2 = 0 in Fig. 7.1). Using q_{ult}, either from Eq. (7.6) or from Eq. (7.7), ultimate load Q_{ult} is determined by multiplying q_{ult} with footing dimensions. Factor of safety (FS) is determined as follows:

$$FS = \frac{Q_{ult}}{P} \quad ...(7.8)$$

There are other considerations in the determination of bearing capacity of soil that will liquefy during design earthquake. Distance of bottom of footing to top of liquefied soil layer is one important consideration. This parameter is difficult to determine for soil that is below groundwater table and has factor of safety against liquefaction that is slightly greater than 1. The reason being, earthquake might induce liquefaction of the upper layer as well. In addition to vertical loads, footing might also be subjected to static and dynamic lateral loads during earthquake. They are dealt with separately. In conventional analysis, vertical load is applied at center of footing. For earthquake loading, footing is often subjected to a moment. This moment is represented by a load having some eccentricity (Meyerhof, 1953). There are standard techniques to determine eccentricity. Eccentrically loaded footing induces higher bearing pressure under one side of footing than on the other side. The largest and the smallest bearing pressures are determined as follows:

$$q' = \frac{Q(B+6e)}{B^2} \quad ...(7.9)$$

$$q'' = \frac{Q(B-6e)}{B^2} \quad ...(7.10)$$

where,
q' = largest bearing pressure under footing.
q'' = smallest bearing pressure under footing.
Q = load per unit length of footing. This includes dead, live and seismic loads acting on footing as well as its self weight.
e = eccentricity of footing.
B = footing width.

It has been suggested that Q should be located within middle one-third of the footing (Eccentricity should be within middle one-third of footing). It has also been suggested that q′ should not exceed allowable soil pressure. These suggestions have been made from safety point of view of the foundation. Factor of safety FS is determined as follows using q′:

$$FS = \frac{q_{ult}}{q'} \quad ...(7.11)$$

q_{ult} in Eq. (7.11) is determined using Eq. (7.6) for strip footing and Eq. (7.7) for spread footing.

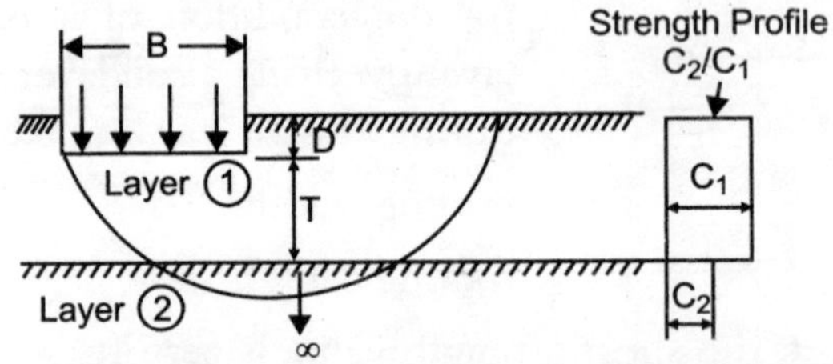

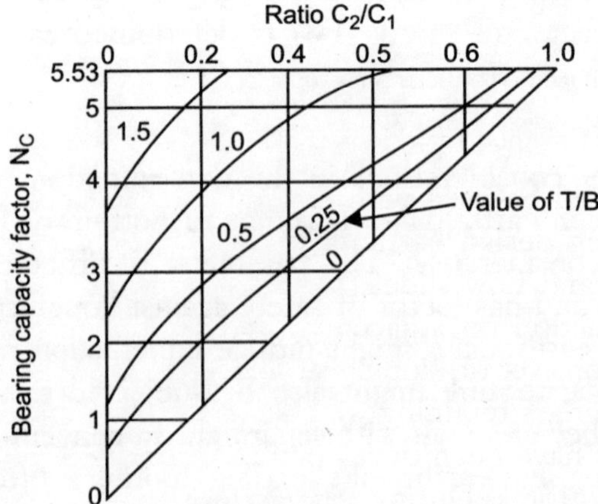

Fig. 7.1 Bearing capacity factor N_c for two layer soil system (Courtesy: Day, 2002)

Calculation of eccentricity and reduction of area is illustrated in Fig. 7.2. Reduction in footing dimension in both dimensions is applicable only for the cases where footing is subjected to moment along long dimension of footing as well as across the footing. If the footing is subjected to moment only in one direction, footing dimension is reduced only in that direction. These reduced footing dimensions are then used to determine bearing capacity. Factor of safety FS is also determined using Fig. 7.2 as follows:

$$FS = \frac{Q_{ult}}{Q} \qquad ...(7.12)$$

Q_{ult} in Eq. (7.12) is determined by multiplying q_{ult} with reduced footing dimensions. q_{ult} is determined using Eq. (7.6) and Eq. (7.7) with reduced footing dimensions.

Special design techniques are available for sloping ground conditions under earthquake loading conditions. Special design techniques are also for inclined base of footing under earthquake loading conditions. Methods have also been developed to determine allowable bearing capacity of foundations at top of slopes.

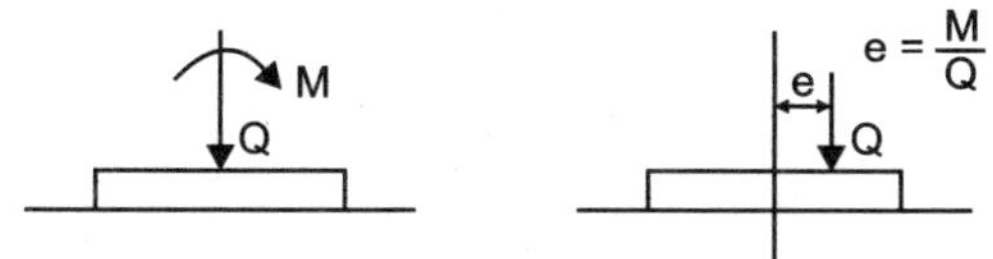

Resultant force acts at the centroid of the reduced area.

(A) Equivalent Loadings

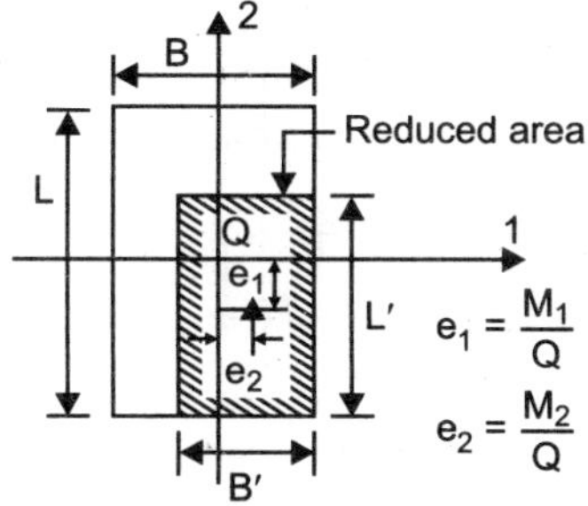

For rectangular footings reduce dimension as follows:

$L' = L - 2e_1 \qquad e_1 = \frac{M_1}{Q}$

$B' = B - 2e_2 \qquad e_2 = \frac{M_2}{Q}$

(B) Reduced area-rectangular footing

Fig. 7.2 Reduced area method for a footing subjected to a moment (Courtesy: Day, 2002)

These methods should be used with caution when dealing with earthquake analysis of soil that will liquefy during design earthquake. The site could be impacted by liquefaction induced lateral spreading and flow slides. Even if the general vicinity of the site is relatively level, the effect of liquefaction on adjacent slopes or retaining walls should be included in the analysis. If the site consists of sloping ground, a slope stability analysis should also be performed. Similarly, if there is retaining wall adjacent to site, a retaining wall analysis should also be performed. Charts have been developed to determine the bearing capacity factors for footings having inclined bottoms. During the earthquake, the inclined footing could translate laterally along the sloping soil or rock contact. If a sloping contact of underlying hard material will be encountered during excavation of footing, the hard material should be excavated in order to construct a level footing that is entirely founded within hard strata.

7.3 GRANULAR SOIL WITH EARTHQUAKE INDUCED PORE WATER PRESSURE

Some times due to earthquake, granular soil doesn't liquefy. Howevere, there is reduction in its shear strength due to increase in pore pressure. This situation is applicable to granular soils below groundwater. Furthermore, factor of safety against liquefaction should be between 1 and 2 for the analysis presented in this subsection.

Following equations are used.

For strip footing: $$q_{ult} = 0.5(1 - r_u)\gamma_b BN_\gamma \quad ...(7.13)$$

For spread footing: $$q_{ult} = 0.4(1 - r_u)\gamma_b BN\gamma \quad ...(7.14)$$

where, r_u = pore water pressure ratio. It is determined by determining factor of safety against liquefaction of soil below footing base. Its value should be in between 1 and 2. When factor of safety against liquefaction is greater than 2, Terzaghi bearing capacity equation can be applied by incorporating the effect of ground water table to determine pore water pressure ratio (Wallace, 1961).

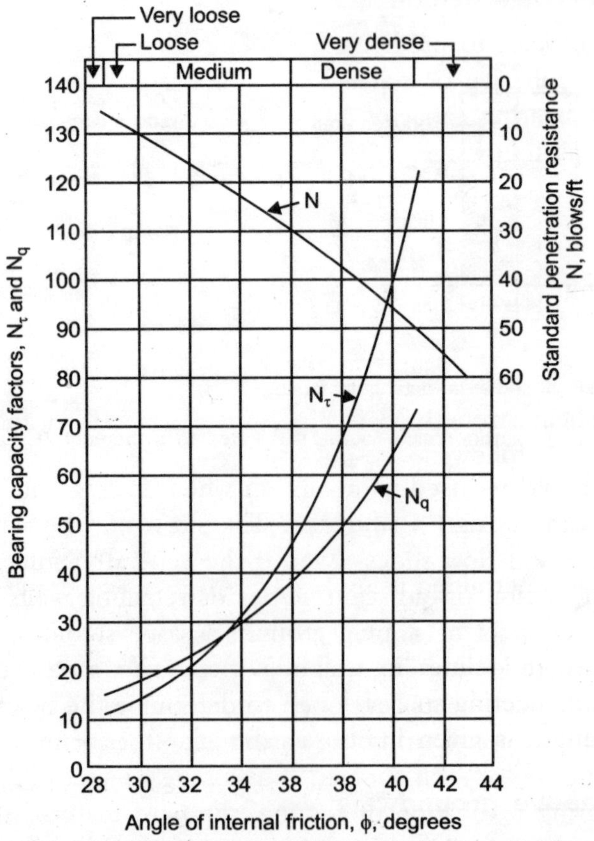

Fig. 7.3 Bearing capacity factors (Courtesy: Day, 2002)

γ_b = buoyant unit weight of soil below footing.

B = footing width.

N_γ = bearing capacity factor based on effective friction angle ϕ' as per Fig. 7.3.

Factor of safety FS is determined as follows:

$$FS = \frac{q_{ult}}{q_{all}} \quad ...(7.15)$$

q_{ult} in Eq. (7.15) is determined using Eq. (7.13) or (7.14). q_{all} in Eq. (7.15) is allowable bearing capacity.

7.4 BEARING CAPACITY ANALYSIS FOR COHESIVE SOIL WEAKENED BY EARTHQUAKE

Cohesive soils as well as organic soils can also be susceptible to a loss of shear strength during the earthquake. In dealing with such soils, it is often desirable to limit the stress exerted by the footing during the earthquake. The stress exerted should be less than the maximum past pressure of the cohesive or organic soils. This prevents the soil from squeezing out. It also prevents soil from deforming laterally from underneath footing.

It is often very difficult to predict the amount of earthquake induced settlement for foundations bearing on cohesive or organic soils. Consequently, in one approach adequate factor of safety against bearing capacity failure of foundation is ensured. Ultimate bearing capacity is determined as follows:

For strip footing:

$$q_{ult} = 5.5s_u \quad ...(7.16)$$

For spread footing:

$$q_{ult} = 5.5s_u\left(1+0.3\frac{B}{L}\right) \quad ...(7.17)$$

s_u is undrained shear strength, B is footing width and L is footing length. Factor of safety FS is determined as follows:

$$FS = \frac{q_{ult}}{q_{all}} \quad ...(7.18)$$

q_{ult} in Eq. (7.18) is determined using Eq. (7.16) or (7.17). q_{all} in Eq. (7.18) is allowable bearing capacity.

There are standard guidelines in terms of undrained shear strength that should be utilized for earthquake engineering analysis (Triandafilidis, 1965). These guidelines for selection of undrained shear strength is given in subsections below.

7.4.1 Cohesive soil above groundwater table

These soils above groundwater table have negative pore pressures. This is due to capillary tension. This tends to hold soil particles together. It also provides additional strength.

Undrained shear strength should be determined by performing unconfined compression or vane shear tests under these conditions. Due to negative pore pressure, a future increase in water content will tend to decrease undrained shear strength of partially saturated cohesive soil. Consequently, possible change in water content in future should also be considered. Ultimate strength obtained during unconfined compression test should be used in bearing capacity analysis.

7.4.2 Cohesive soil below groundwater table having low sensitivity

Sensitivity is ratio of undrained shear strength of undisturbed soil to undrained shear strength of completely remolded soil. Consequently, it represents loss of strength as cohesive soil is remolded. Earthquake also tends to shear the soil back-forth. Furthermore, it also remolds it. For low sensitivity soils (sensitivity < 4), reduction of undrained shear strength during earthquake is small. Consequently, undrained shear strength from unconfined compression or vane shear test should be used for bearing capacity analysis.

7.4.3 Cohesive soil below groundwater table having high sensitivity

For high sensitivity soils (sensitivity > 8), earthquake-induced ground shaking could lead to significant shear strength loss during earthquake shaking. The stress-strain curve from an unconfined compression test on such soils exhibits peak shear strength developed at low vertical strain. This is followed by dramatic drop-off in strength with continued straining. Estimated reduction in undrained shear strength due to earthquake shaking should be included in analysis. Most critical condition develops when such soil is subjected to high static shear stress. If sum of static shear stress and seismic induced shear stress during earthquake shaking exceeds undrained shear strength, there is significant reduction in shear strength (Cunny and Sloan, 1961). Cohesive soils having sensitivity in between 4 and 8 tend to be intermediate case.

There are other factors also which may be considered in bearing capacity analysis. Peak ground acceleration and earthquake magnitude is such factor. Higher the peak ground acceleration and higher the magnitude of earthquake, greater the tendency for cohesive soil to be strained and remolded by earthquake shaking. Undrained shear strength, sensitivity, maximum past pressure and stress-strain behavior are other important soil behavior parameters, which should be included in the analysis. Increase in shear stress due to dynamic loading should also be included in analysis. Lightly loaded foundations tend to produce small dynamic loads. On the other hand heavy and tall buildings subject foundation to high dynamic loads due to rocking. Finally it can be concluded that since so many variables are involved, it takes considerable judgement in selection of undrained shear strength to be used in Equations (7.16) and (7.17).

Finally, based on results of settlement analysis and bearing capacity analysis for both static and dynamic conditions design recommendations such as minimum footing dimensions, embedment requirements and allowable bearing capacity values are provided. Consequently, the objective of earthquake resistant design of shallow foundations is achieved to support varied civil engineering structures.

Example 7.1. At a particular site, ground surface is horizontal and the zone of liquefaction extends from a depth of 1.2 m to 6.7 m. During construction, additional 1.8 m thick cohesive soil is placed at ground surface. After that it is proposed to construct a sewage disposal plant at the site. Bottom of the footing for the plant is to be at a depth of 0.5 m below ground surface. For both existing 1.2 m thick unliquefiable cohesive soil layer and additional 1.8 m thick cohesive layer, the undrained shear strength is 60 kPa. Calculate the factor of safety of the footing using punching shear analysis for:

(*a*) 1m wide strip footing under total load of 60 kN/m.

(*b*) 2m wide square spread footing under total load of 600 kN.

Solution:

(*a*) For strip footing, using Eq. (7.1),

$T = 1.8 + 1.2 - 0.5 = 2.5$ m,

τ_f = undrained shear strength of cohesive soil = 60 kPa.

$P = 60$ kN/m.

Substituting the values in Eq. (7.1):

$FS = 5.0$

(*b*) For square spread footing, using Eq. (7.2),

$T = 1.8 + 1.2 - 0.5 = 2.5$ m,

τ_f = undrained shear strength of cohesive soil = 60 kPa,

$P = 600$ kN

and $B = L = 2$ m.

Substituting the values in Eq. (7.2):

$FS = 2.0$

Example 7.2. Perform total stress analysis using Terzaghi equations for general and local shear failure to find out factor of safety for 1m wide strip footing. Use data from Example 7.1.

Solution: From Example 7.1, P = 60 kN/m for 1 m wide strip footing, $T = 1.8 + 1.2 - 0.5 = 2.5$ m. $c_1 = s_u = 60$ kPa = 60 kN/m^2 & $c_2 = 0$. i.e. $T/B = 2.5/1 = 2.5$ and $c_2/c_1 = 0$. For these two values and using Fig. 7.1, Nc = 5.5. Consequently, using Eq. (7.6), $q_{ult} = (60)(5.5) = 330$ kN/m^2 for strip footing. Hence Q_{ult} for 1 m wide strip footing = q_{ult} B = (330)(1) = 330 kN/m. Using Eq. (7.8), factor of safety = FS = 5.5.

Example 7.3. Use data from Example 7.1. Assume that apart from vertical loads, the strip and the spread footing is subjected to earthquake induced moment equal to 5 kN.m/m and 150 kN.m which act in single (B) direction. Determine factor of safety using Eq. (7.11).

Solution:

(*a*) For 1 m wide strip footing, Q = P = 60 kN/m, e = M/Q = 5/60 = 0.0833 m, for middle one-third of footing, e can not exceed 0.17 m, and therefore e is within middle one-third of footing. $q' = 89.988$ kN/m^2 from Eq. (7.9). q_{ult} in Eq. (7.11) is determined using Eq. (7.6) which makes use of Fig. 7.1. Hence $q_{ult} = 330$ kN/m^2. Consequently, Factor of safety = FS = 3.667.

(b) For 2 m wide square spread footing, Q = P = 600 kN, e = M/Q = 150/600 = 0.25 m. For middle one-third of footing, e can not exceed 0.33 m, and therefore e is within middle one-third of footing. Q = 600/2 = 300 kN/m for use in Eq. (7.9). $q' = 262.5$ kN/m^2 from Eq. (7.9) for 2m wide square spread footing. Furthermore, T = 1.8 + 1.2 – 0.5 = 2.5m. $c_2 = 0$ and $c_1 = 60$ kN/m^2. T/B = 2.5/2 = 1.25m and $c_2/c_1 = 0$. Using these and from Fig. 7.1, $N_c = 3.2$. Hence $q_{ult} = 249.6$ kN/m^2 from Eq. (7.7) with B = L = 2 m. Consequently, Factor of safety = FS = 0.95.

Home Work Problems

1. Solve Example 7.1 assuming that both the existing 1.2 m thick and additional 1.8m thick unliquefiable soil layer is cohesionless with effective friction angle equal to 31°. Coefficient of earth pressure at rest is equal to 0.5. Total unit weight of soil above water table is 18.3 kN/m^3 and buoyant unit weight of soil below water table is 9.7 kN/m^3. Water table is at a depth of 1.2m below existing ground surface. (**Ans.** (a) FS = 0.8 (b) FS = 0.32)

2. Perform total stress analysis using Terzaghi equations for general and local shear failure to find out factor of safety for 2 m wide square spread footing. Use data from Example 7.1. (**Ans.** FS = 1.664)

3. Use data from Example 7.1. Assume that apart from vertical loads, the strip and the spread footing is subjected to earthquake induced moment equal to 5 kN.m/m and 150 kN.m which act in single (B) direction. Determine factor of safety using Eq. (7.12) (**Ans.** (a) FS = 4.58 (b) FS = 1.176.

4. A site consists of a sand deposit with a fluctuating groundwater table. The expected depth of footing will be 0.5 to 1 m. Assume that groundwater table can rise to a level close to footing base. Buoyant unit weight of sand is 9.65 kN/m^3, effective friction angle for sand = 32° and pore water pressure ratio = 0.2. Using factor of safety of 5, determine allowable bearing capacity for:

   (*a*) 1.5m wide strip footing.

   (*b*) 2.5m wide square spread footing.

   (**Ans.** (*a*) 24.318 kPa (*b*) 32.424 kPa))

5. What are the guidelines to calculate undrained shear strength in the bearing capacity analysis for cohesive soil weakened by earthquake?

EARTHQUAKE RESISTANT DESIGN OF DEEP FOUNDATION

8.1 INTRODUCTION

Deep foundations are used when the upper soil stratum is too soft, weak or compressible to support the static and earthquake induced foundation loads. Deep foundations are also used when there is possibility of undermining of the foundation, either in static or earthquake induced foundation loading condition. One example is bridge pier which is often founded on deep foundation to prevent a loss of support due to flood conditions which could cause river bottom scour. Furthermore, in the case of excessive settlement, there is bearing capacity failure due to liquefaction of underlying soil deposit as well as ground surface damage during earthquake. To prevent consequent structural damage, deep foundations are used.

The most common types of deep foundations are piles and piers supporting individual footing or mat foundations. Piles are relatively long, slender, columnlike members. They are often made up of steel, concrete or wood. Either they are driven in or cast in place in predrilled holes. There are different types of piles. Batter piles are driven at an angle inclined to vertical. This provides high resistance to lateral loads. If the soil liquefies during earthquake, lateral resistance of batter pile may be significantly reduced. End-bearing pile is another type of pile. For end-bearing piles, the support capacity of pile is derived principally from the resistance of foundation material on which pile tip rests. End-bearing piles are used when a soft layer is underlain by dense or hard stratum. If upper soft layer liquefies during earthquake, the pile will be subjected to down drag forces. Consequently, pile must be designed to resist these soil-induced forces. In friction piles, support capacity of pile is derived principally from the resistance of soil friction and/or adhesion mobilized along side of pile. They are used in soft clays where the end bearing resistance is small due to punching shear at pile tip. If the soil is subjected to liquefaction during earthquake, both the frictional resistance and lateral resistance of pile may be lost during earthquake. Combined end bearing and friction piles are another type of piles. These piles derive its support capacity from combined end bearing resistance developed at pile tip and friction and/or adhesion resistance on pile perimeter.

Pier is defined as a deep foundation system. It is similar to cast in place pile. Pier consists of a column like reinforced concrete member. Piers have often large enough diameter to enable down hole inspection. They are also referred to as drilled shafts, bored piles or drilled caissons.

There are some more techniques available for forming deep foundation elements resistant to earthquake. Mixed in place soil cement or soil lime piles are called mixed in place piles. Vibroflotation is another method used to make a cylindrical, vertical hole. This hole is filled with compacted open graded gravel or crushed rock. These stone columns also have the additional capacity of reducing the potential for soil liquefaction. This is achieved by allowing earthquake induced pore water pressures to dissipate rapidly. The pore water flows into the highly permeable open-graded gravel or crushed rock. They are also called vibroflotation-replacement stone columns. Grouted stone columns are also used. In grouted stone columns, voids are filled up with bentonite-cement or water-sand-bentonite cement mixtures. Concrete vibroflotation column is also used as deep foundation element. In concrete vibroflotation columns, concrete is used instead of gravel to fill the hole. All these special types of pile foundations are used as earthquake resistant piles.

8.2 DESIGN CRITERIA

Different items are used in designing and construction of piles which can resist earthquake induced loads. They are given in subsections below.

8.2.1 Engineering Analysis

Based on the results of engineering analysis, it has been suggested that deep foundation should be designed and constructed such that it penetrates all the soil layers that are expected to liquefy during earthquake. In these cases, deep foundations derive support from unliquefiable soil located below potentially troublesome soil strata which is prone to liquefaction. Possibility of down drag forces as well as loss of lateral resistance due to soil liquefaction should be incorporated in the analysis. If a liquefiable soil layer is located below bottom of deep foundation, then punching shear analysis should be used because there is possibility of deep foundation's punching into underlying soil strata. This analysis has already been explained in the context of shallow foundations. For end-bearing piles, load applied to pile cap can be assumed to be transferred to pile tip. Based on shear strength of unliquefiable soil below bottom of piles as well as vertical distance from pile tip to liquefiable soil layer, factor of safety can be calculated using Equations (7.1) and (7.2). B and L represents width and length respectively of pile group.

8.2.2 Field Load Tests

Prior to foundation construction, a pile or pier should be load tested in field. Testing is necessary to determine its carrying capacity. There are uncertainities involved in engineering analysis of pile design. Consequently, pile load tests are recommended. Pile load test result in more economical foundation than those based solely on engineering analysis. These tests are useful to evaluate dynamic loading conditions as well. The test method is used to provide data on strain of pile under impact load. Force, acceleration, velocity and displacement of a pile under impact load is also obtained from these tests. These data are used to estimate bearing capacity and integrity of pile. Hammer performance, pile stresses and soil dynamic characteristics are also obtained from these data. However, field load tests can't simulate response of pile for situations where soil is expected to liquefy during design earthquake. Consequently, results of pile load tests would have to be modified for the expected liquefaction conditions.

8.2.3 Application of Pile Driving Resistance

Initially the pile capacity was estimated based on driving resistance. Driving resistance was obtained during installation of pile. Pile driving equations were developed. They are called Engineering News Formula. These equations relate the pile capacity to the energy of pile driving hammer as well as the average net penetration of pile per blow of the pile hammer (White, 1964). However, no satisfactory relationship between pile capacity from pile driving equations and pile capacity measured from load tests have been observed. It has been concluded that pile driving equations are no longer justified (Terzaghi and Peck, 1967). Furthermore, for high displacement piles that are closely spaced, the vibration and soil displacement associated with pile driving densifies granular soil around the pile. Consequently, the liquefaction resistance of soil is increased due to pile driving.

8.2.4 Specifications and Experience

Other factors included in earthquake resistant deep foundation design include governing building code, agency requirement and local experience. Local experience, such as deep foundation performance during prior earthquakes, can be a important factor in the design and construction of pile foundations.

Home Work Problems

1. What are the design criteria for earthquake resistant design of deep foundations?

SLOPE STABILITY ANALYSES FOR EARTHQUAKES

9.1 INTRODUCTION

Slope movement is secondary effect of earthquake. There can be many types of earthquake induced slope movement. For rock slopes, earthquake induced slope movement is divided into falls and slides. Falls have relatively free falling nature of rock or rocks due to earthquake. In slides, there is shear displacement along a distinct failure surface due to earthquake. Falls and slides occur in soil slopes also. In addition, slope can be subjected to flow slide or lateral spreading also during earthquake. For a specific type of earthquake induced slope movement to occur, minimum slope inclination is required which ranges from 40° in earthquake induced rock fall to 0.3° in liquefaction induced lateral spreading.

For seismic evaluation of slope stability, analysis can be grouped in two general categories:

1. Inertia slope stability analysis
2. Weakening slope stability analysis

Inertia slope stability analysis is preferred if material retains its shear strength during earthquake. Pseudostatic and Newmark are two common methods in this analysis. Weakening slope stability analysis is preferred if material experiences significant shear strength reduction during earthquake. During liquefaction, there are two cases of weakening slope stability analyses. Flow slide develops when the static driving forces exceed shear strength of soil along failure surface. In lateral spreading static driving forces do not exceed shear strength of soil along slip surface. Instead, driving forces only exceed resisting forces during those portions of earthquake that impart net inertial forces in the downward direction. This results in progressive and incremental lateral movement.

Massive crystalline bedrock and sedimentary rock (retaining intact during earthquake), soils which dilate during seismic shaking, soils not exhibiting reduction in shear strength with strain, clay with low sensitivity, soils located above water table and landslides having distinct rupture surface are examples where material retain shear strength during earthquake. Inertia slope analyses is preferred for them.

Foliated or friable rock which fractures during earthquake, sensitive clays, overloaded soft and organic soils as well as loose soils located below water table and under liquefaction induced excess pore water pressure are examples where material experience sufficient shear strength reduction during earthquake. Weakening slope stability analyses is preferred for them.

9.2 INERTIA SLOPE STABILITY-PSEUDOSTATIC METHOD

This method is easy to understand and is applicable for both total and effective stress slope stability analyses. The method ignores cyclic nature of earthquake. It assumes that additional static force is applied on the slope due to earthquake. In actual analysis, a lateral force acting through centroid of sliding mass is applied which acts in out of slope direction. This pseudostatic lateral force F_h is calculated as follows:

$$F_h = ma = \frac{Wa}{g} = \frac{Wa_{max}}{g} = k_h W \quad ...(9.1)$$

where, F_h = horizontal pseudostatic force acting through centroid of sliding mass in out of slope direction. For two dimensional analysis, slope is usually assumed to have unit length.

m = total mass of slide material.

W = total weight of slide mass.

a = acceleration, maximum horizontal acceleration at ground surface due to earthquake. ($= a_{max}$)

a_{max} = peak ground acceleration.

a_{max}/g = seismic coefficient.

Earthquake subjects sliding mass in general to vertical as well as horizontal pseudostatic forces. Since vertical pseudostatic force on sliding mass has very little effect on its stability, it is ignored.

Based on the results of field exploration and laboratory testing, unit weight of soil or rock can be determined. Consequently, weight of sliding mass, W can be readily calculated. On the other hand, selection of seismic coefficient takes considerable experience and judgement. Certain guidelines regarding selection of seismic coefficient is as follows:

1. Higher the value of peak ground acceleration, higher the value of k_h.
2. k_h is also determined as function of earthquake magnitude.
3. When items 1 and 2 are considered, k_h should never be greater than a_{max}/g.
4. Sometimes local agencies suggest minimum value of seismic coefficient.
5. For small slide mass, $k_h = a_{max}/g$

6. For intermediate slide mass, $k_h = 0.65a_{max}/g$
7. For large slide mass, $k_h = 0.1$ for sites near faults generating 6.5 magnitude earthquake and , $k_h = 0.15$ for sites near faults generating 8.5 magnitude earthquake.
8. $k_h = 0.1$ for severe earthquake, = 0.2 for violent and destructive earthquake and = 0.5 for catastrophic earthquake.

9.2.1 Wedge Method

This is simplest type of slope stability analysis (refer Fig. 9.1). Failure wedge has planar slip surface, inclined at an angle α to horizontal. Analysis could be performed for the case of planar slip surface intersecting the face of slope or passing through toe of slope.

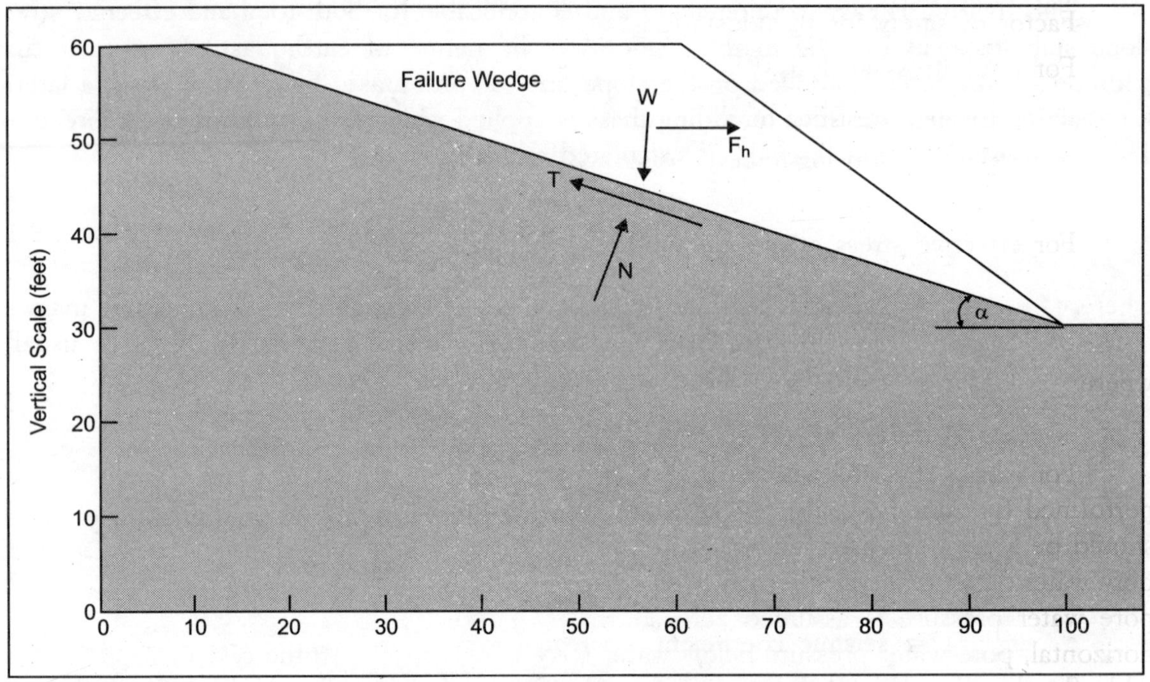

Fig. 9.1 Wedge method (Courtesy: Day, 2002)

As per pseudostatic wedge analysis of Fig. 9.1, four forces are acting:

W = weight of failure wedge = total unit weight γ_t times cross-sectional area of failure wedge for assumed unit length of slope.

$F_h = k_h W$ = horizontal pseudostatic force acting through centroid of sliding mass in out of slope direction.

N = normal force acting on slip surface.

T = shear force acting along slip surface.

For total stress analysis:

$$T = cL + N\tan\phi = s_u L$$

For effective stress analysis:

$$T = c'L + N'\tan\phi'$$

where,

L = length of planar slip surface

c, ϕ = shear strength parameters for total stress analysis

s_u = undrained shear strength of soil for total stress analysis

N = total normal force acting on slip surface

$c'\phi'$ = shear strength parameters for effective stress analysis

N' = effective normal force acting on slip surface

Factor of safety for pseudostatic analysis is obtained as follows:

For total stress analysis:

$$FS = \frac{\text{resisting force}}{\text{driving forces}} = \frac{cL + N\tan\phi}{W\sin\alpha + F_h\cos\alpha} = \frac{cL + (W\cos\alpha - F_h\sin\alpha)\tan\phi}{W\sin\alpha + F_h\cos\alpha} \quad \text{...(9.2a)}$$

For effective stress analysis:

$$FS = \frac{c'L + N'\tan\phi'}{W\sin\alpha + F_h\cos\alpha} = \frac{c'L + (W\cos\alpha - F_h\sin\alpha - uL)\tan\phi'}{W\sin\alpha + F_h\cos\alpha} \quad \text{...(9.2b)}$$

where, FS = factor of safety for pseudostatic analysis

u = average pore water pressure along slip surface

For total stress analysis, total stress parameters of soil should be known and is often performed for cohesive soils. For effective stress analysis, effective stress parameters of soil should be known and is often performed for cohesionless soils. For effective stress analysis, pore water pressure along slip surface should also be known. For soil layers above water table, pore water pressure is assumed zero. If the soil is below water table and water table is horizontal, pore water pressure below water table is hydrostatic. In the case of sloping water table flow net can be used to estimate pore water pressure below water table.

9.2.2 Method of Slices

In this method, failure mass is subdivided into vertical slices and factor of safety is determined based on force equilibrium equations. A circular arc slip surface and rotational type of failure mode is often used in this method.

The resisting and the driving forces are calculated for each slice and then summed to obtain factor of safety of the slope. The equation to calculate factor of safety is identical to Eq. (9.2), with driving and resisting forces calculated for each slice and then summed to obtain factor of safety. However, there are more unknowns than equilibrium equations in the method of slices. Consequently, an assumption is to be made concerning interslice forces. In ordinary method of slices, resultant of interslice forces is parallel to average inclination of slice, α. Bishop simplified, Janbu simplified, Janbu generalized, Spencer method and Morgenstern-

Price method are other methods of slices. Because of the tedious nature of calculations, computer programs are routinely used to perform the pseudostatic slope stability analysis using the method of slices. It has not been discussed in detail in this book.

9.2.3 Other Slope Stability Considerations

Important factors which are needed in the cross section to be used for pseudostatic slope stability analysis is as follows:

Different soil layers: If the slope contains different soil or rock type, with different engineering properties, it must be incorporated in the analysis. For all soil layers, either effective shear strength or shear strength in terms of total stress parameters must be known. Horizontal pseudostatic force is specified for every layer.

Slip surfaces: Either planar or composite type slip surface may be needed for analysis.

Tension cracks: Tension cracks at the top of slope can reduce factor of safety of a slope by as much as 20 percent. This should be included in the analysis. Destabilizing effects of water in tension cracks should also be included in the analysis.

Surcharge loads: Surcharge loads (at top or even on slope face) as well as tie-back anchors should be included in the analysis.

Nonlinear shear strength envelope: If shear strength envelope of soil is non linear, it should be included in the analysis.

Plane strain condition: Long uniform slopes are plane strain condition. Friction angle in this case is about 10% higher than the friction angle obtained in triaxial experiment. This should be included in the analysis.

These considerations are incorporated in Eq. (9.2) to complete the analysis as per actual conditions.

9.3 INERTIA SLOPE STABILITY – NEWMARK METHOD

Purpose of this method is to estimate the slope deformation for those cases where the pseudostatic factor of safety is less than 1.0, which corresponds to failure condition. It is assumed that slope will deform during those portions of earthquake when out of slope earthquake forces make pseudostatic factor of safety below 1.0 and the slope accelerates downwards. Longer the duration for which pseudostatic factor of safety is zero, greater the slope deformation.

Fig. 9.2(a) shows horizontal acceleration of slope during earthquake. Accelerations plotting above zero line are out of slope and accelerations plotting below zero line are into slope accelerations. Only out of slope accelerations cause downslope movement and are used in the analysis. a_y in Fig. 9.2(a), is horizontal yield acceleration and corresponds to pseudostatic factor of safety exactly equal to 1. Portion of acceleration pulses above a_y (darkened portion in Fig. 9.2(a)), causes lateral movement of slope. Fig. 9.2(b) and (c) represent horizontal velocity and slope displacement due to darkened portion of acceleration pulse. Slope displacement is incremental and occurs only when horizontal acceleration due to earthquake exceeds a_y.

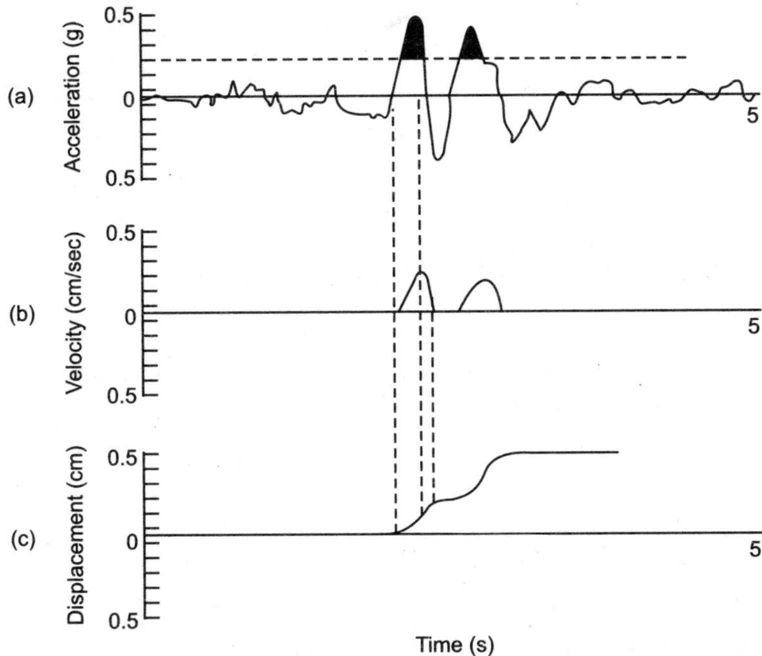

Fig. 9.2 Diagram illustrating Newmark method (a) acceleration versus time (b) velocity versus time for darkened portion of acceleration pulse (c) corresponding downslope displacement versus time in response to velocity pulses (Courtesy: Day, 2002)

Magnitude of slope displacement depends on variety of factors. Higher the a_y value, more stable the slope is for a given earthquake. Greater the difference between peak ground acceleration a_{max} due to earthquake and a_y, larger the downslope movement. Longer the earthquake acceleration exceeds a_y, larger the downslope deformation. Larger the number of acceleration pulses exceeding a_y, greater the cumulative downslope movement during earthquake. Most common method used in Newmark method is as follows:

$$\log d = 0.90 + \log\left[\left(1-\frac{a_y}{a_{max}}\right)^{2.53}\left(\frac{a_y}{a_{max}}\right)^{-1.09}\right] \quad \text{...(9.3)}$$

where, d = estimated downslope movement due to earthquake in cm.

a_y = yield acceleration.

a_{max} = peak ground acceleration of design earthquake.

Essentially a_{max} must be greater than a_y. While using Eq. (9.3), pseudostatic factor of safety is determined first using the technique described in Fig. 9.2. If it is less than 1, k_h is reduced till pseudostatic factor becomes equal to 1. This value of k_h is used to determine a_y using Eq. (9.1). This a_y and a_{max} is used to determine slope deformation. Analysis is more accurate for small and medium size failure masses.

9.3.1 Limitations of Newmark Method

Major assumption of Newmark method is that the slope will deform only when peak ground acceleration exceeds yield acceleration. Analysis is most appropriate for wedge type failure.

One limitation of Newmark method is that it is unreliable for slopes not deforming as single massive block. Slope composed of dry and loose granular soil is such slope. Earthquake induced settlement of dry and loose granular soils depend on relative density, maximum shear strain induced by earthquake and number of shear strain cycles. It is anticipated that the lateral movement of slope is the same order of magnitude as the calculated settlement.

9.4 WEAKENING SLOPE STABILITY-FLOW SLIDES

Weakening slope stability is preferred for materials which experience significant reduction in shear strength during earthquake. Analysis is done for flow slides in this section. Flow slides develop when static driving forces exceed weakened shear strength of soil along slip surface. Consequently, factor of safety is less than 1. There are three types of flow slides.

Mass liquefaction occurs when nearly the entire sloping mass is susceptible to liquefaction. They occur to partially or fully submerged slopes. First step of analysis is to determine factor of safety against liquefaction. If the entire sloping mass or a significant part of it is subjected to liquefaction during earthquake, slope will be susceptible to flow slide.

Zonal liquefaction occurs when there is specific zone of liquefaction within the slope. First step is to determine the location of zone of soil expected to liquefy during design earthquake. Slope stability analysis is performed using circular arc slip surfaces passing through zone of expected liquefaction. If factor of safety of slope is less than 1, flow slide is likely to occur during earthquake.

Landslide movement due to soil liquefaction occurs due to liquefaction of horizontal soil layers. There could be liquefaction of layers of saturated soil within the slope. This can cause entire slope to move laterally along liquefied layer at base. Potential liquefiable soil layer may be thin, hard to discover during subsurface exploration and hence it is difficult to evaluate landslide movement possibility due to earthquake. Since slip surface must pass through these horizontal layers, slope stability analysis is often performed using block type failure mode.

9.4.1 Factor of Safety Against Liquefaction for Slopes

First step is to determine zones likely to liquefy due to earthquake and to determine factor of safety against liquefaction. For level ground it can be determined using analysis presented in Chapter 6. This factor of safety thus obtained should be adjusted for sloping ground conditions. This is done using chart of Fig. 9.3.

In Fig. 9.3, horizontal axis is α, defined as:

$$\alpha = \frac{\tau_{h\ static}}{\sigma'_{vo}} \qquad ...(9.4)$$

where, $\tau_{h\ static}$ = static shear force acting on horizontal plane.

σ'_{vo} = vertical effective stress.

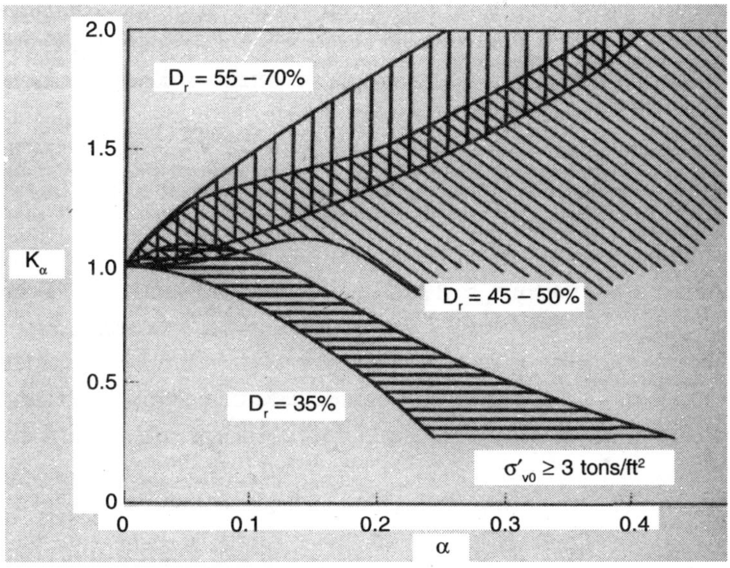

Fig. 9.3 Chart for use to adjust factor of safety against liquefaction for sloping ground (Courtesy: Day, 2002)

For infinite slopes, α is approximately equal to slope ratio (= vertical distance/horizontal distance). Vertical axis of Fig. 9.3 is K_α. To determine factor of safety against liquefaction for sloping ground, factor of safety against liquefaction obtained from Chapter 6 is multiplied with K_α of Fig. 9.3. D_r in Fig. 9.3 represents relative density of soil.

There are some approximate suggested guidelines. For $\alpha \leq 0.10$, use $K_\alpha = 1.0$. For $D_r \geq 45\%$, use $K_\alpha = 1.0$. For $\alpha > 0.10$ and $D_r < 45\%$, K_α is determined from Fig. 9.3 which requires considerable experience and judgement.

9.4.2 Stability Analysis for Liquefied Soil

Factor of safety against liquefaction based on level ground surface is determined first at various soil depths. Then this factor of safety is adjusted for sloping ground conditions using Fig. 9.3. If the entire soil depth, or significant portion of it will be subjected to liquefaction, then the slope will be susceptible to flow slides.

For the case of zonal liquefaction, slope stability analysis is required for soil that is likely to liquefy during earthquake. There are two different approaches. In the first approach, pore water pressure ratio ($= u/(\gamma_t h)$) of liquified soil is taken as 1. u is pore water pressure, γ_t = total unit weight of soil, and h = depth below ground surface. Pore water pressure ratio 1 means pore water pressure is equal to total stress and hence effective stress is zero. This approach is used with effective stress analysis and when effective cohesion of soil is zero. If soil doesn't liquefy during earthquake, effective shear strength parameters and estimated pore water pressures are used in slope stability analysis. Second approach assumes liquefied soil to have zero shear strength. For total stress analysis, undrained shear strength (s_u) is zero and for effective stress analysis effective stress parameters are zero. However, shear strength of liquefied soil may not necessarily be equal to zero. This undrained liquefied shear strength is termed as liquefied shear strength. It has been found to be correlated with $(N_1)_{60}$ value. But, since undrained liquefied shear strength is very small, most conservative slope stability analysis is performed for flow slides using effective stress analysis assuming liquefied shear strength equal to zero. Further details are beyond the scope of this book.

9.5 WEAKENING SLOPE STABILITY-LIQUEFACTION INDUCED LATERAL SPREADING

If the liquefaction induced lateral spreading is restricted to localized ground surface, it is called localized lateral spreading. If it causes lateral movement over an extensive distance, it is called large scale lateral spreading. Large scale lateral spreading has been discussed in detail. Concept of cyclic mobility is used to describe large scale lateral spreading. The driving forces only exceed resisting forces during those portions of earthquake that impart net inertial force in downward direction. Each cycle of net inertial force causes driving forces to exceed resisting forces resulting in progressive and incremental lateral movement. Usually the ground surface first cracks at unconfined toe and then ground cracks progressively move upslope.

Amount of horizontal ground displacement resulting from liquefaction induced lateral spreading is determined using empirical methods. They have been developed based on regression analysis. These equations are as follows:

For lateral spreading towards free face (river bank for example):

$$\log D_H = -16.366 + 1.178M - 0.927\log R - 0.013R + 0.657\log W + 0.348\log T + 4.527\log(100\text{-}F) - 0.922D_{50} \quad ...(9.5)$$

For lateral spreading of gently sloping ground:

$$\log D_H = -15.787 + 1.178M - 0.927\log R - 0.013R + 0.429\log S + 0.348\log T + 4.527\log(100\text{-}F) - 0.922D_{50} \quad ...(9.6)$$

where,

D_H = horizontal ground displacement due to lateral spreading, meters.

M = earthquake magnitude of design earthquake.

R = distance to expected epicenter or nearest fault rupture of design earthquake in km.

W = free face ratio, expressed as percentage, = 100H/L. H is height of free face and L is horizontal distance from base of free face to site location.

T = cumulative thickness (meters) of submerged sand layers having $(N_1)_{60} < 15$.

F = fines content of soil comprising layer T, expressed as percentage. It is percent of soil particles based on dry weight that pass No. 200 sieve.

D_{50} = grain size corresponding to 50 percent fines of soil comprising layer T, mm.

S = slope gradient (vertical/horizontal), expressed as percentage.

Furthermore, it has been reported that sites subjected to $M \leq 8$ earthquake and have soils with $(N_1)_{60}$ values > 15 are resistant to lateral spreading. Equations 9.5 and 9.6 need not be applied. Equations 9.5 and 9.6 are accurate within a factor of ± 2 and D_H from these equations should be multiplied by 2 for conservative design estimate of lateral spreading.

To obtain reliable deformations, terms in Equations (9.5) and (9.6) must be within following ranges:

$6 \leq M \leq 8$

$1\% < W < 20\%$

$0.1\% \leq S \leq 6\%$

$1m \leq T \leq 15m$

$F \leq 50\%$

$D50 \leq 1mm$

Other limitations of Equations (9.5) and (9.6) are:

(i) Liquefied soil layer must be within 10m of ground surface.

(ii) Equations (9.5) and (9.6) overestimate displacement due to lateral spreading of liquefied gravels.

(iii) Equation (9.5) should be applied with caution at sites very close to free face.

(iv) For free face, both Equations (9.5) and (9.6) should be used. Higher value should be used in actual design.

9.5.1 Summary

The liquefaction of soil can cause flow failure or lateral spreading. Even with factor of safety against liquefaction greater than 1, there could still be significant weakening of soil and deformation of slope. To summerize:

1. For factor of safety against liquefaction $\leq$ 1, soil is expected to liquefy due to earthquake. Flow slide analysis (sec. 9.4) and/or lateral spreading analysis (sec. 9.5) will be performed.
2. For factor of safety against liquefaction > 2, the pore water pressure due to earthquake is usually small. It can be neglected. Soil is not weakened by earthquake and inertia slope stability analysis (sec. 9.2 and sec. 9.3) will be performed.
3. For factor of safety against liquefaction greater than 1 and less than or equal to 2, soil is not expected to liquefy due to earthquake. However, there could be substantial pore water pressure increase. Pore water pressure ratio can be estimated as a function of factor of safety against liquefaction. Using this pore water pressure ratio, effective stress slope stability analysis could be performed. If analysis shows factor of safety less than 1, failure of slope during earthquake is expected.

Example 9.1:

A slope has a height of 9.1 m and the slope face is inclined at 2:1 (horizontal:vertical). Assume wedge type analysis, where slip surface is planer through toe of slope and is inclined at 3:1 (horizontal:vertical). Total unit weight of slope material = 18.1 kN/m^3. Using undrained shear strength parameters of c = 14.5 kPa and $\phi = 0$, calculate factor of safety for static case and for earthquake condition of $k_h = 0.3$. Assume that it is not a weakening type soil.

Solution:

Refer Fig. 9.1, for the information given in the problem, area of the wedge = 0.5(9.1)(27.3 – 18.2) = 41.4m^2. For unit length of slope, total weight of wedge, W = (41.4)(18.1) = 750 kN/m.

Static case:

$$F_h = 0$$

Using Eq. (9.2(a)) and the information given in the problem:

$c = 14.5$ kN/m^2, $\phi = 0$, $\alpha = \tan^{-1}1/3 = 18°$ and $L = 9.1/\sin\alpha = 9.1/\sin 18 = 29$ m. Substituting the values in Eq. (9.2(a)):

$$FS = \frac{(14.5)(29)}{(750)(\sin 18)} = 1.8$$

Earthquake case:

$$F_h = 0.3\ W.$$

Other values are same as static case. Substituting the values in Eq. (9.2(a)):

$$FS = \frac{(14.5)(29)}{(750)(\sin 18) + (0.3)(750)(\cos 18)} = 0.94$$

Example 9.2:

Use data from Example 9.1. Calculate slope deformation based on Newmark method. Peak ground acceleration $a_{max} = 0.3$ g.

Solution:

Since pseudostatic factor of safety is less than 1, slope deformation based on Newmark method can be estimated. From Eq. 9.2(a), for FS =1, k_h comes out to be 0.26. Hence a_y = 0.26 g. a_{max} = 0.3 g, given. Substituting in Eq. (9.3),

$$\log d = -1.25.\ \text{So}\ d = 0.06\text{cm}$$

Example 9.3:

A slope is inclined at an angle of 14°. Relative density of soil comprising slope is 35%. Factor of safety against liquefaction for level ground is 1.25. Find out factor of safety against liquefaction for sloping ground. Assume slope to be infinite.

Solution:

For infinite slope, $\alpha = \tan 14 = 0.25$. Also, relative density of soil = 35%. From Fig. 9.3, $K_\alpha = 0.5$. Hence, factor of safety for sloping ground condition = (1.25)(0.5) = 0.625.

Home Work Problems

1. Use data from Example 9.1, except assume that slip surface has effective shear strength of $c' = 4$ kPa and $\phi' = 29°$. Average measured steady state pore water pressure u = 2.4kPa. Determine factor of safety of failure wedge based on effective stress analysis for static condition and earthquake condition of $k_h = 0.2$. It is not weakening type soil and pore pressure will not increase due to earthquake. (Ans. static FS = 2.04, earthquake FS = 1.194)
2. Use data from problem 1. Calculate slope deformation based on Newmark method. Peak ground acceleration = 0.3 g. (**Ans.** 0)
3. Using empirical method to predict amount of horizontal ground displacement resulting from liquefaction induced lateral spreading, determine horizontal ground displacement due to lateral spreading for following condition:

   * free face condition
   * factor of safety against flow slide > 1
   * M = 7.5
   * R = 50km
   * W = 10%
   * S = 5%
   * T = 5m
   * F = 6%
   * D_{50} = 0.38mm

   (**Ans.** 1.8m)
4. Explain about different types of flow slides.
5. Differentiate between inertia and weakening slope stability. Give examples for each.

CHAPTER 10

RETAINING WALL ANALYSES FOR EARTHQUAKES

10.1 INTRODUCTION

Retaining wall is a structure whose primary purpose is to provide lateral support for soil or rock. It may also support vertical loads. They could be of gravity, cantilever, counterfort and crib wall type. Basement walls and bridge abutments are typical examples. Performance of retaining wall during earthquake is very complex. Due to seismic forces, walls can move by translation and/or rotation depending on wall design. Magnitude and distribution of dynamic wall pressure is influenced by mode of wall movement. Maximum soil thrust acts on wall when the wall translates or rotates towards the backfill. It is minimum when the wall translates or rotates away from the backfill. The shape of earthquake pressure distribution and the point of application of resultant changes as the wall moves. Dynamic wall pressure and permanent wall displacement increase significantly, near the natural frequency of wall backfill system under earthquake loading. Increased residual pressure may remain on wall after episode of strong shaking has ended.

It has been stated that the allowable bearing pressure and allowable passive pressure should be increased by a factor of one-third while performing seismic analysis. This increase is appropriate if retaining wall bearing material and soil in front of wall consists of massive crystalline bedrock and sedimentary rock that remains static during earthquake, soils which dilate due to earthquake, soils having little reduction in shear strength with strain, clay with low sensitivity and soils located above water table. However the increase is not recommended if the soil consists of foliated rock that fractures in earthquake, loose soil located below water table, sensitive clays and soft clays. Former group of soils do not loose shear strength during seismic shaking while later group of soils loose shear strength during seismic shaking.

This chapter deals with methods of retaining wall analysis under earthquakes.

10.2 PSEUDOSTATIC METHOD

This method is easy to understand and apply. The method ignores cyclic nature of earthquake and treats as if it is applying additional static force on retaining wall. Pseudostatic approach is to apply a lateral force upon retaining wall. This lateral force acts through the centroid of active wedge. The active wedge is zone of soil involved in the development of active earth pressure on the wall. It is inclined at an angle of $45^0 + \phi/2$ from horizontal, as indicated in Fig. 10.1. ϕ is angle of internal friction of soil.

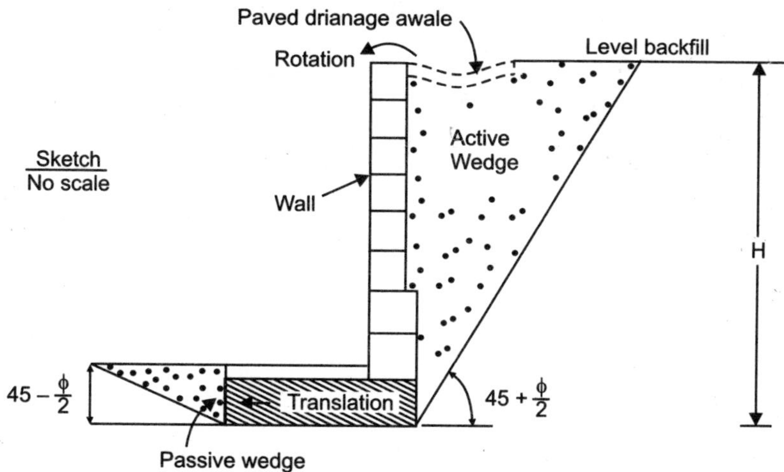

Fig. 10.1 Active wedge behind retaining wall (Courtesy: Day, 2002)

Pseudostatic lateral force P_E is calculated by the following equation:

$$P_E = ma = \frac{W}{g}a = W\frac{a_{max}}{g} = k_h W \quad ...(10.1)$$

where, P_E = horizontal pseudostatic force acting on retaining wall. Wall is assumed to have unit length and this force acts through centroid of active wedge.

m = total mass of active wedge.

W = total weight of active wedge.

a = acceleration.

a_{max} = peak ground acceleration.

$a_{max}/g = k_h$ = seismic coefficient (pseudostatic coefficient).

Earthquake subjects active wedge to both vertical and horizontal pseudostatic forces. But vertical force is ignored since it has small effect on retaining wall design. k_h is assumed to be a_{max}/g. From Fig. 10.1,

$$W = \frac{1}{2}HL\gamma_t = \frac{1}{2}H[H\tan(45° - \phi/2)]\gamma_t = \frac{1}{2}k_A^{1/2}H^2\gamma_t \quad ...(10.2)$$

where, W = weight of active wedge, per unit length of wall.

H = height of retaining wall.

L = length of active wedge at top of retaining wall.

γ_t = total unit weight of backfill soil.

k_A = active earth pressure coefficient. Often the wall friction is neglected.

Substituting Eq. 10.2 in Eq. 10.1,

$$P_E = k_h W = \frac{1}{2} k_h k_A^{1/2} H^2 \gamma_t = \frac{1}{2} k_A^{1/2} \left(\frac{a_{max}}{g}\right)(H^2\gamma_t) \quad ...(10.3)$$

Since P_E acts to the centroid of active wedge, location of P_E is at a distance of $\frac{2}{3}H$ above the base of retaining wall. According to Seed and Whitman (1970),

$$P_E = \frac{3}{8}\frac{a_{max}}{g}H^2\gamma_t \quad ...(10.4)$$

Location of P_E is at a distance of 0.6 H above wall base. According to Mononobe-Okabe method,

$$P_{AE} = P_A + P_E = \frac{1}{2}k_{AE}H^2\gamma_t \quad ...(10.5)$$

where, P_{AE} = sum of static (P_A) and pseudostatic earthquake force (P_E). Equation for k_{AE} is shown in Fig. 10.2. In Fig. 10.2, ψ is defined as,

$$\psi = \tan^{-1}k_h = \tan^{-1}\frac{a_{max}}{g} \quad ....(10.6)$$

Force P_{AE} acts at a distance of $\frac{1}{3}H$ above wall base. Retaining wall is further analyzed for sliding and for overturning. Factor of safety for sliding using pseudostatic as well as using Seed and Whitman analysis is given as,

$$FS = \frac{N\tan\delta_1 + P_P}{P_H + P_E} \quad ...(10.7)$$

where, N = Sum of weight of wall, footing and vertical component of active earth pressure resultant force. Vertical component of active earth pressure resultant force = $P_A\sin\delta$. $P_A = 0.5k_A\gamma_t H^2$. k_A is obtained from equation in Fig. 10.2. H is height of retaining wall and γ_t is unit weight of backfill soil. δ_1 is friction between bottom of foundation and soil backfill. $P_H = P_A\cos\delta$. δ is friction between back face of wall and soil back fill. P_P is passive resistance force divided by reduction factor which is taken as 2. Usually, the wall friction and slight slope of the front of retaining wall is neglected in the calculation of P_P. P_E is obtained from Eq. 10.3 for pseudostatic and from Eq. 10.4 for Seed and Whitman analysis. Factor of safety for sliding using Mononobe-Okabe method is given as,

$$FS = \frac{N\tan\delta_1 + P_P}{P_H} \qquad \text{...(10.8)}$$

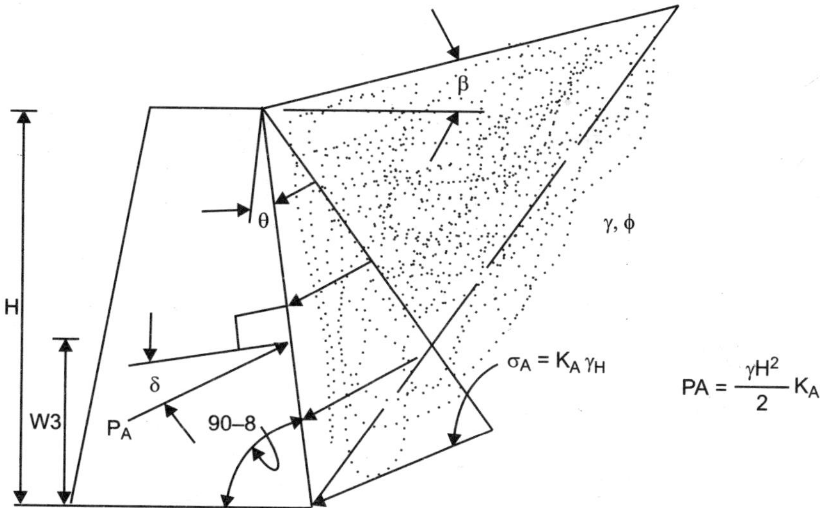

Fig. 10.2 k_A equation for static and k_{AE} equation for earthquake condition (Courtesy, Day, 2002)

(A) Coulomb's Equation (Static Condition):

$$K_A = \frac{\cos^2(\phi-\theta)}{\cos^2\theta\cos(\delta+\theta)\left[1+\sqrt{\dfrac{\sin(\delta+\phi)\sin(\phi-\beta)}{\cos(\delta+\theta)\cos(\beta-\theta)}}\right]^2}$$

(B) Mononobe-Okabe Equation (Earthquake Condition):

$$K_{AE} = \frac{\cos^2(\phi-\theta-\psi)}{\cos\psi\cos^2\theta\cos(\delta+\theta+\psi)\left[1+\sqrt{\dfrac{\sin(\delta+\phi)\sin(\phi-\beta-\psi)}{\cos(\delta+\theta+\psi)\cos(\beta-\theta)}}\right]^2}$$

Where, N = Sum of weight of wall, footing + $P_{AE}\sin\delta$. P_{AE} is obtained from Eq. 10.5. $P_H = P_{AE}\cos\delta$. Method of obtaining P_P is same as in Eq. 10.7. Factor of safety for overturning using pseudostatic as well as using Seed and Whitman analysis is given as,

$$FS = \frac{Wa}{0.333P_H H - Pve + 0.667HP_E} \qquad \text{...(10.9)}$$

Where, a = lateral distance from resultant weight W of wall and footing to toe of footing. $P_H = P_A\cos\delta$ and $P_v = P_A\sin\delta$. P_A determination has been explained in the context of Eq. 10.7. e = lateral distance from P_v to the toe of wall. Factor of safety for overturning using Mononobe-Okabe method is given as,

$$FS = \frac{Wa}{0.333HP_{AE}\cos\delta - eP_{AE}\sin\delta} \quad ...(10.10)$$

Where, a = lateral distance from resultant weight W of wall and footing to toe of footing. e = lateral distance from P_v to the toe of wall. P_{AE} determination has been explained in the context of Eq. 10.8. Under combined static and earthquake loads, factor of safety for sliding as well as for overturning should be in the range of 1.1 to 1.2.

10.3 RETAINING WALL ANALYSIS FOR LIQUEFIED SOIL

There are three types of liquefaction damages. Firstly, there is liquefaction in front of retaining wall. This reduces passive resistance in front of retaining wall. Secondly, soil behind the retaining wall liquefies, and pressure exerted on wall is greatly increased. These two effects can work individually or together causing sliding, overturning or tilting failure. Thirdly, there could be liquefaction below bottom of wall causing bearing capacity failure.

10.3.1 Design Pressures

Firstly, adjusted factor of safety against liquefaction for soil behind retaining wall, front of retaining wall and from below the bottom of soil is calculated using analysis presented in Chapter 6.

For soils subjected to liquefaction in paasive zone, liquified soil is assumed zero shear strength. Consequently, it doesn't provide sliding or overturning resistance.

For soils subjected to liquefaction in active zone, pressure exerted on face of wall increases. Zero shear strength of liquefied soil is assumed. If water level is located only behind retaining wall, thrust on wall due to liquefaction of backfill is calculated with k_A = 1 and $\gamma_t = \gamma_{sat}$. If water levels are approximately the same on both sides of retaining wall, k_A = 1 and $\gamma_t = \gamma_{sub}$.

For liquefaction of bearing soil, use analysis presented in Sec. 7.2.

10.3.2 Sheet Pile Walls

In Fig. 10.3, term D represents portion of sheet pile anchored in soil. H represents unsupported face of sheet pile wall. Ap represents restraining force on sheet pile wall due to tieback construction. At the groundwater table (point A),

Active earth pressure at $A = k_A \gamma_t d_1$ ...(10.11)

Where, k_A = active earth pressure coefficient neglecting friction, γ_t = total unit weight of soil above water table, d_1 = depth from ground surface to groundwater table.

At point B, active earth pressure equals,

Active earth pressure at $B = k_A \gamma_t d_1 + k_A \gamma_b d_2$ ...(10.12)

Where, γ_b = buoyant unit weight of soil below water table, d_2 = depth from groundwater table to bottom of sheet pile wall.

At point C, passive earth pressure is given as,

Passive earth pressure at $C = k_p \gamma_b D$ ...(10.13)

Where, k_p is passive earth pressure coefficient neglecting friction.

Static design of sheet pile wall requires following analysis:

(i) evaluation of earth pressures that acts on wall.

(ii) determination of required depth D of piling penetration.

(iii) calculation of maximum bending moment M_{max}.

(iv) selection of appropriate pile type.

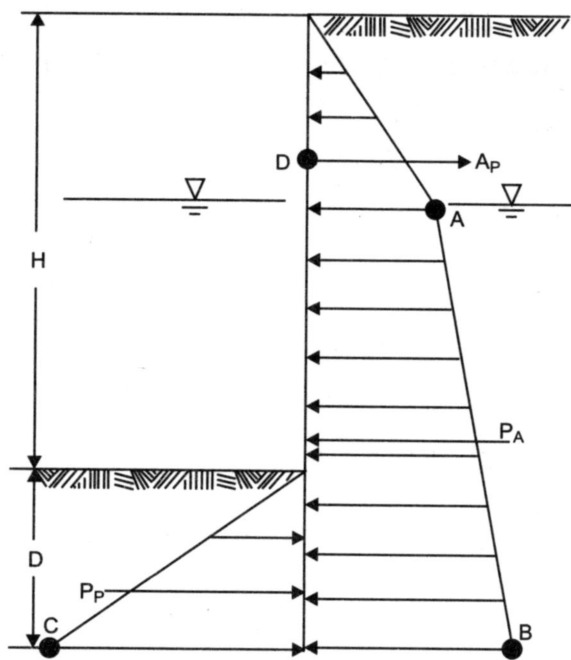

Fig. 10.3 Static design of sheet pile wall (Courtesy, Day, 2002)

Typical design process is to assume depth D and calculate factor of safety for toe failure. Factor of safety is defined as moment due to passive force (resisting moment) divided by moment due to active force (destabilizing moment) at the tieback anchor (point D). This value should be between 2 and 3. Once D has been calculated, Anchor pull Ap can be calculated using,

$$A_P = P_A - \frac{P_P}{FS} \quad ...(10.14)$$

P_A and P_p are resultant active and passive forces. FS is factor of safety and is obtained as ratio of moment due to passive force to moment due to active force. Other design aspects for static analysis have not been discussed in this book.

In the case of factor of safety against liquefaction (greater than 2) for sand behind, below and front of sheet pile, due to earthquake, there will be horizontal pseudostatic force acting on sheet pile. It will be acting at a height of 0.667(H+D) from sheet pile bottom. Since water pressure tends to cancel on both sides of wall, the pseudostatic force is calculated using Eq. 10.3 based on buoyant unit weight. Moment due to psudostatic force at the tieback anchor will have the same direction as moment due to active force. Incorporating this, factor of safety is calculated for earthquake condition. Anchor pull is obtained by adding P_E in Eq. 10.14 for earthquake condition with FS being factor of safety for earthquake condition. Partial passive wedge liquefaction due to earthquake in the case of sheet pile walls has not been discussed in this book. For liquefaction of entire active wedge due to earthquake and no liquefaction of soil in front of sheet pile wall, moment due to passive force at tieback anchor will be unaltered. Lateral pressure due to liquefied soil is determined with $k_A = 1$ and submerged unit weight of liquefied soil. Ratio of moments due to passive force to moment due to lateral force of liquified soil at tieback anchor is used to find out factor of safety in this case.

10.4 RETAINING WALL ANALYSES FOR WEAKENED SOIL

If only backfill soil is weakened due to earthquake, the force exerted on the back face of wall increases. Shear strength corresponding to weakened condition of backfill soil is calculated and is used to determine forces exerted on wall. Using this, bearing pressure, factor of safety for sliding, factor of safety for overturning and location of resultant vertical force is calculated.

If soil beneath bottom of wall or soil in passive wedge is weakened due to earthquake, there could be additional settlement, bearing capacity failure, sliding failure or overturning failure. Weakening of ground beneath or in front of wall could result in shear failure beneath retaining wall. Design approach is to reduce shear strength of bearing soil or passive wedge soil to account for its weakened state during earthquake. Settlement, bearing capacity, factor of safety for sliding, overturning and shear failure beneath the bottom of wall is calculated for this weakened soil.

If there is weakening of backfill soil and reduction in soil resistance, combined analysis of previous two conditions is done.

10.5 RESTRAINED RETAINING WALLS

In these walls, movement of retaining wall is restricted. Static earth pressure at rest is determined using coefficient of earth pressure at rest k_0 using conventional technique of static earth pressure calculation. For earthquake conditions, restrained retaining wall is subjected to larger forces compared to retaining walls having the ability to develop active wedge. Pseudostatic method is as follows,

$$P_{ER} = \frac{P_E k_0}{k_A} \tag{10.15}$$

where, P_{ER} = pseudostatic force acting on restrained retaining wall.

P_E = pseudostatic force assuming wall has ability to develop active wedge using Eq. 10.3, 10.4 or 10.5.

k_0 = coefficient of earth pressure at rest.

k_A = active earth pressure coefficient obtained from Fig. 10.2.

10.6 TEMPORARY RETAINING WALLS

Static design of temporary braced walls is shown in Fig. 10.4. If the sand deposit has groundwater table above the level of bottom of excavation, water pressure must be added to the case 'a' pressure distribution of Fig. 10.4. Since excavations are temporary, undrained shear strength (s_u = c) should be used in the analysis in clays (cases 'b' and 'c'). Pressure distribution of case 'b' and 'c' is not valid for permenent wall or for walls where water table is above bottom of excavation.

Earthquake design is done using technique described in sec. 10.2 or sec. 10.5 based on whether wall is considered yielding or restrained. Weakening of soil during earthquake and its effect on temporary retaining wall should also be included in the analysis.

Example 10.1:

Refer Fig. 10.1. Assume H = 4m, thickness of reinforced concrete wall stem = 0.4m and reinforced concrete wall footing is 3m wide by 0.5m thick. Ground surface in front of wall is level with top of wall footing and unit weight of concrete = 25 kN/m^3. Wall backfill consists of sand having ϕ = 32° and γ_t = 20 kN/m^3. Sand in front of wall has same properties. Friction angle between bottom of footing and bearing soil, δ_1 = 38°. For level backfill and neglecting wall friction on back side of wall and front side of footing, determine:

(i) resultant normal force.

(ii) factor of safety for sliding.

(iii) factor of safety for overturning.

For static condition using pseudostatic analysis and for earthquake conditions using Eq. 10.3 if a_{max} = 0.20g.

Solution:

Static condition:

Resultant normal force = Sum of weight of wall, footing and vertical component of active earth pressure resultant force.

But, vertical component of active earth pressure resultant force = $P_v = P_A \sin\delta$. In this problem δ = 0 as there is no friction between backfill soil and wall face.

Hence, resultant normal force = Sum of weight of wall and footing = (3.5)(0.4)(25)+(3)(0.5)(25) = 35 + 37.5 = 72.5 kN/m.

Factor of safety for sliding = FS = $\dfrac{N \tan\delta_1 + P_P}{P_H + P_E}$, with N = 72.5 kN/m, δ_1 = 38°, P_P = $0.5\, k_p \gamma_t H^2$ ($k_p = \tan^2(45 + \phi/2) = 3.25$) divided by reduction factor (2) = (0.5)(3.25)(20)(0.5)^2 divided by reduction factor (2) = 8.125 kN/m divided by reduction factor (2) = 4.06 kN/m.

$P_H = P_A\cos\delta = 0.5\gamma_t H^2 k_A \cos\delta$. k_A will be obtained from static equation of Fig. 10.2 with $\theta = \beta = \delta = 0$ according to this problem. Hence k_A = $(1-\sin\phi)/(1+\sin\phi)$ = 0.307. So, $P_H = P_A$ = (0.5)(20)(4)^2(0.307) = 49.12 kN/m and P_E = 0 for static case. Substituting the values, factor of safety for sliding = $\dfrac{(72.5)(\tan 38) + (4.06)}{49.12} = 1.236$

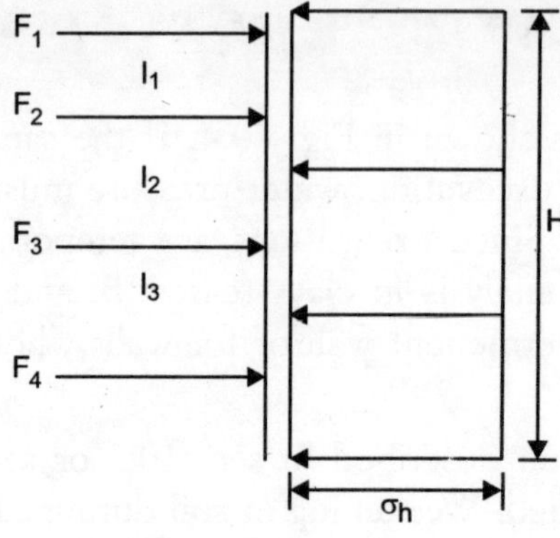

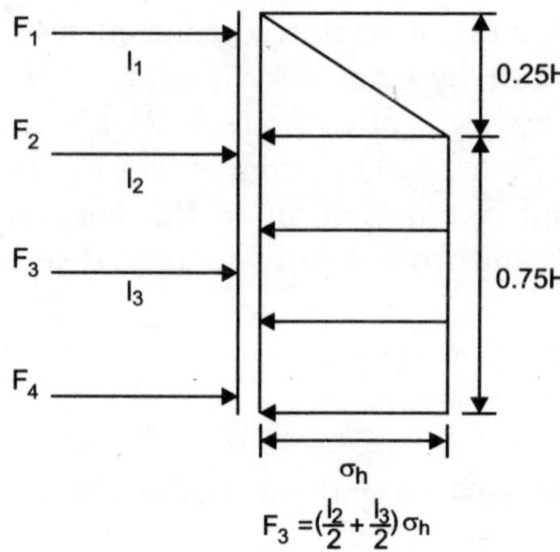

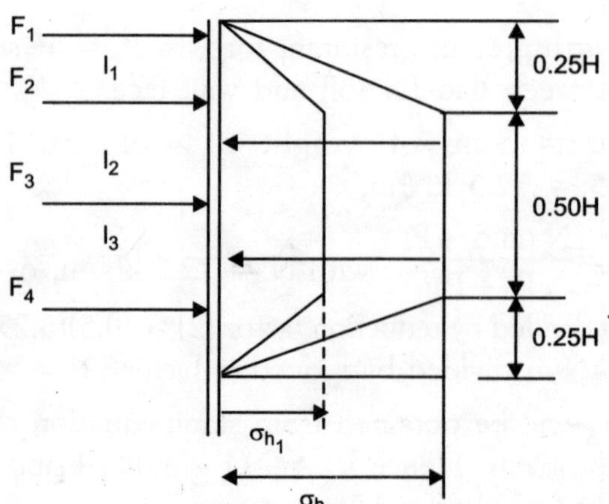

Fig. 10.4 Earth pressure distribution on temporary braced walls (Courtesy, Day, 2002)

$$\text{Factor of safety for overturning} = \frac{W_a}{0.333P_H H - P_v e + 0.667HP_E} = \frac{(35)(2.8)+(37.5)(1.5)}{(0.333)(49.12)(4)}$$

= 2.35 (P_v = 0, explained above and P_E = 0 for static case).

Earthquake condition:

Resultant normal force will be same as before = 72.5 kN/m.

Using Eq. 10.3, $P_E = \frac{1}{2}k_A^{1/2}\left(\frac{a_{max}}{g}\right)(H^2\gamma_t) = (0.5)(0.307)^{0.5}(0.2)(4)2(20) = 17.7$ kN/m.

$$\text{Factor of safety for sliding} = \frac{N\tan\delta_1 + P_p}{P_H + P_E} = \frac{(72.5)(\tan 38)+(4.06)}{(49.12)+(17.7)} = 0.908$$

$$\text{Factor of safety for overturning} = \frac{W_a}{0.333P_H H - P_v e + 0.667HP_E}$$

$$= \frac{(35)(2.8)+(37.5)(1.5)}{(0.333)(49.12)(4)+(0.667)(4)(17.7)} = 1.37$$

Example 10.2

Refer mechanically stabilized earth retaining wall shown in Fig. 10.5. Let H = 20ft, width of mechanically stabilized retaining wall = 14ft, depth of embedment at front of stabilized zone = 3ft. Soil behind and in front of stabilized zone is clean sand with ϕ = 30° at total unit weight of 110 lb/ft³. There is no friction along vertical back and front side of mechanically stabilized zone. For mechanically stabilized zone, soil has total unit weight = 120 lb/ft³ and δ_1 = 23° along bottom of mechanically stabilized zone. Calculate resultant normal force. Also calculate factor of safety for sliding and overturning under static conditions.

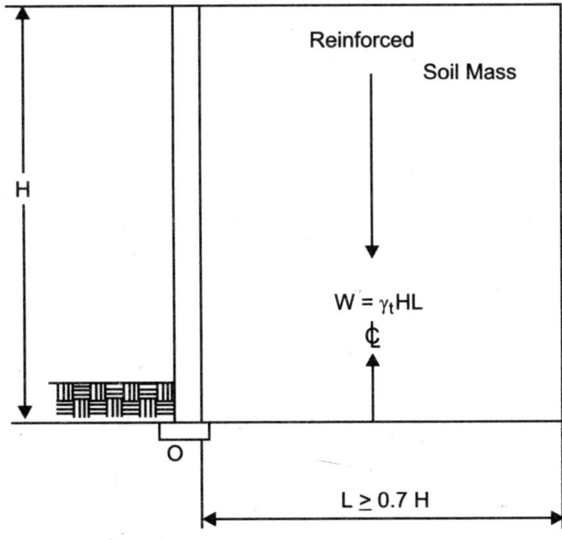

Fig. 10.5 Mechanically stabilized earth retaining wall.

Solution:

Resultant normal force = N = $HL\gamma_t$ = (20)(14)(120) = 33600 lb/ft. Factor of safety for sliding = $\dfrac{N\tan\delta_1 + P_P}{P_H + P_E}$

N = 33600 lb/ft, δ_1 = 23° given, P_E = 0 for static condition, $P_H = P_A = 0.5k_A\gamma_t H^2$. $k_A = \tan^2(45-\phi/2) = \tan^2(45-30/2) = 0.333$, γ_t = 110 lb/ft³, H = 20ft. Hence, P_H = 7326 lb/ft. Passive force = $0.5k_P\gamma_t D^2$ = (0.5)(1/0.333)(110)(3)² = 1486.49 lb/ft. P_P = passive force/reduction factor = 1486.49/2 = 743.245 lb/ft. Hence factor of safety for sliding = 2.05

$$\text{Factor of safety for overturning} = \frac{W_a}{0.333P_H H - P_v e + 0.667HP_E}$$

$$= \frac{(33600)(7)}{(0.333)(7326)(20)} = 4.82$$

Example 10.3

Refer Fig. 10.3. Soil behind and front of sheet pile is uniform sand with ϕ' = 33°, γ_b = 64 lb/ft³ and γ_t = 130 lb/ft³. H = 30ft and D = 20ft. Water level in front of wall and groundwater table is at same elevation at 5ft below ground surface. Tieback anchor is located at 4ft below ground surface. Neglecting wall friction, determine factor of safety and tieback anchor force for static and earthquake condition using pseudostatic method for a_{max} = 0.20g.

Solution:

Static case:

$$k_A = \tan^2(45-\phi/2) = \tan^2(45-16.5) = 0.295,$$
$$kp = 1/k_A = 3.39.$$

From 0ft to 5ft,

$$P_{1A} = 0.5k_A\gamma_t(5)^2 = (0.5)(0.295)(130)(5)^2$$
$$= 479.375 \text{ lb/ft}$$

From 5ft to 50ft,

$$P_{2A} = k_A\gamma_t(5)(45) + (0.5)(k_A)(\gamma_b)(45)^2$$
$$= (0.295)(130)(5)(45) + (0.5)(0.295)(64)(45)^2$$
$$= 8628.75 + 19116 = 27744.75 \text{ lb/ft.}$$
$$P_A = P_{1A} + P_{2A} = 479.375 + 27744.75 = 28224.125 \text{ lb/ft}$$
$$P_P = (0.5)(k_P)(\gamma_b)(D)^2 = (0.5)(3.39)(64)(20)^2$$
$$= 43392 \text{ lb/ft}$$

Moment due to passive force at tieback anchor = (43392)(26 + (0.667)(20)) = 1.707×10^6

Neglecting P_{1A}, moment due to active force at tieback anchor = (8628.75)(1+45/2) + (19116)(1 + (0.667)(45)) = (2.02775×10^5)+(5.9288×10^5) = 7.95655×10^5.

$$\text{Factor of safety} = \frac{1.707 \times 10^6}{7.95655 \times 10^5} = 2.145$$

$$\text{Anchor pull} = A_P = P_A - \frac{P_p}{FS} = 28224.125-(43392/2.145) = 7994.75 \text{ lb/ft}$$

Earthquake case:

$P_E = \frac{1}{2} k_A^{1/2} \left(\frac{a_{max}}{g}\right)(H^2\gamma_b) = (0.5)(0.295)^{0.5}(0.2)(50)^2(64) = 8690$ lb/ft acting at 0.667(H + D) from bottom of sheet pile wall.

Moment due to P_E at tieback anchor = 8690[(0.333)(50) – (4)] = 1.10×10^5

Total destabilizing moment = $7.95655 \times 10^5 + 1.10 \times 10^5 = 9.05655 \times 10^5$

$$\text{Factor of safety} = \frac{1.707 \times 10^6}{9.05655 \text{ x } 10^5} = 1.884$$

$$\text{Anchor pull} = A_P = P_A - \frac{P_p}{FS} + P_E = 28224.125 - \frac{43392}{1.884} + 8690 = 13882.277 \text{ lb/ft}$$

Example 10.4

A braced excavation will be used to support vertical sides of 20ft deep excavation (H = 20ft in Fig. 10.4). If site is sand with $\phi = 33°$ and γ_t = 125 lb/ft³, calculate σ_h and resultant earth pressure force on braced excavation for static and earthquake condition with a_{max} = 0.20g using Eq. 10.3. Ground water is below bottom of excavation.

Solution:

Static case:

$k_A = \tan^2(45-\phi/2) = 0.294$, from Fig. 10.4, $\sigma_h = (0.65)(k_A)(\gamma)(H) = (0.65)(0.294)(125)(20)$ = 477.75 lb/ft²

resultant force = $\sigma_h H$ = (477.75)(20) = 9555 lb/ft

Earthquake case:

$$P_E = \frac{1}{2} k_A^{1/2} \left(\frac{a_{max}}{g}\right)(H^2\gamma_t) = (0.5)(0.294)^{0.5}(0.2)(20)^2(125) = 2711.088 \text{ lb/ft}$$

Home Work Problems

1. Solve Example 10.1 for earthquake condition using Eq. 10.4. (Ans. Resultant normal force = 72.5 kN/m, Factor of safety for sliding = 0.83, factor of safety for overturning = 1.19)
2. Solve Example 10.2 to determine factor of safety for sliding and overturning under earthquake condition with a_{max} = 0.20g. Use Eq. (10.3). (Ans. Factor of safety for sliding = 1.52, factor of safety for overturning = 2.84)
3. Solve Example 10.3 for liquefaction of entire active wedge due to earthquake and no liquefaction of soil in front of sheet pile wall to determine factor of safety. (Ans. FS = 0.726)
4. Solve Example 10.4 for soft clay at site having cohesion as 300 lb/ft^2. Use Eq. 10.4 for earthquake condition. (Ans. Resultant force = 22750 lb/ft, earthquake analysis, P_E = 3750 lb/ft)
5. Explain about retaining wall analysis for weakened soil.
6. Explain about restrained retaining wall design.

EARTHQUAKE RESISTANT DESIGN OF BUILDINGS

11.1 INTRODUCTION

The primary objective of earthquake resistant design is to prevent building collapse during earthquakes. It also minimises the risk of death or injury to people in or around those buildings. Earthquake forces are generated by the inertia of buildings. Inertia of buildings dynamically respond to ground motion. The dynamic nature of the response makes earthquake loadings markedly different from other building loads. Designer temptation to consider earthquakes as 'a very strong wind' is a trap that must be avoided since the dynamic characteristics of the building are fundamental to the structural response and thus the earthquake induced actions are able to be mitigated by design.

The concept of dynamic considerations of buildings is one which sometimes generates unease and uncertainty within the designer. Effective earthquake design methodologies can be, and usually are, easily simplified without detracting from the effectiveness of the design. High level of uncertainty relating to the ground motion generated by earthquakes seldom justifies the often used complex analysis techniques as well as the high level of design sophistication often employed. A good earthquake engineering design is one where the designer takes control of the building by dictating how the building is to respond. This can be achieved by selection of the preferred response mode and selecting zones where inelastic deformations are acceptable. This can also be achieved by suppressing the development of undesirable response modes which could lead to building collapse.

Modern earthquake design has its genesis in the 1920's and 1930's. At that time earthquake design typically involved the application of 10% of the building weight as a lateral force on the structure. This lateral force was applied uniformly up the height of the building. It was not until the 1960's that strong ground motion accelerographs became more available. These instruments record the ground motion generated by earthquakes. When used in conjunction with strong motion recording devices (which were able to be installed at different levels within buildings themselves), it became possible to measure and understand the dynamic response of buildings when they were subjected to real earthquake induced ground motion.

By using actual earthquake motion records as input to, then, recently developed inelastic integrated time history analysis packages, it became apparent that many buildings designed to earlier codes had inadequate strength to withstand design level earthquakes without experiencing significant damage. However, observations of the in-service behaviour of buildings showed that this lack of strength did not necessarily result in building failure or even severe damage when they were subjected to severe earthquake attack. Provided the strength could be maintained without excessive degradation as inelastic deformations developed, buildings generally survived the earthquake. Conversely, buildings which experienced significant strength loss frequently became unstable and often collapsed during earthquakes.

With this knowledge the design emphasis moved to ensure that the retention of post-elastic strength was the primary parameter which enabled buildings to survive the earthquake. It also became clear that some post-elastic response mechanisms were preferable to others. Preferred mechanisms could be easily detailed to accommodate the large inelastic deformations expected. Other mechanisms were highly susceptible to rapid degradation with. Those mechanisms needed to be suppressed. The key to successful modern earthquake engineering design lies therefore in the detailing of the structural elements. Consequently, the desirable post-elastic mechanisms are identified and promoted. On the other hand the formation of undesirable response modes are precluded.

Desirable mechanisms are those which are sufficiently strong to resist normal imposed actions without damage. At the same time, they are capable of accommodating substantial inelastic deformation without significant loss of strength or load carrying capacity. Such mechanisms have been found to generally involve the flexural response of reinforced concrete and steel structural elements or the flexural steel dowel response of timber connectors. Undesirable post-elastic response mechanisms within specific structural elements have brittle characteristics. They include shear failure within reinforced concrete, reinforcing bar bond failures, loss of axial load carrying capacity or buckling of compression members such as columns. They also include the tensile failure of brittle components such as timber or under-reinforced concrete.

11.2 EARTHQUAKE RESISTING PERFORMANCE EXPECTATION

The seismic structural performance requirements of buildings are often prescribed within national building codes. For instance Clause B1 'Structure' of the New Zealand Building Code (New Zealand Government Print, 1992) prescribes that the building is to retain its amenity when subjected to frequent events of moderate intensity earthquake. Furthermore, it is to remain stable and avoid collapse during rare events of high intensity earthquake. The Building Code of Australia (Australian Building Codes Board. 1996) prescribes the performance expectations in similar rather vague terms. It is left to the Loadings Standards of New Zealand and Australia to interpret 'moderate' and 'high' loading intensities. This they do by equating the 'amenity' retention as the **Serviceability Limit State** and collapse avoidance as the **Ultimate Limit State** loads or combinations of loads. Consequently, for compliance with the mandatory provisions of the national building codes the following requirements need to be satisfied:

(i) For amenity retentions (Serviceability Limit State): The building response should remain predominantly elastic. Some minor damage would be acceptable provided any such damage does not require repair. Buildings should remain fully operational. Preservation of the appropriate levels of lateral deformation to protect non-structural damage is of primary importance. The loading intensity for this limit state is to be relatively low.

(ii) For collapse avoidance (Ultimate or Survival Limit State): The risk to life safety is maintained at acceptably low levels. Building collapse is to be avoided. Significant residual deformation is expected within the buildings. Both structural and non-structural members experience damage. Building repair may not be economical. The loading intensity used for design can be equated to rare earthquakes with long return periods. This is the single most important design criterion since it relates to preservation of life. It demands that the system possess adequate overall structural ductility. This enables load redistribution while avoiding collapse.

11.3 KEY MATERIAL PARAMETERS FOR EFFECTIVE EARTHQUAKE RESISTANT DESIGN

Compliance with the performance criteria of the various limit states outlined in previous section requires different material properties. The serviceability limit states criteria demand that certain stiffness and elastic strength parameters be met. This is primarily concerned with the linear stress/strain deformation relationships associated with elastic system response. The ultimate limit state criteria generally demand that an appropriate level of post-elastic ductility capacity is available. This helps to avoid collapse. There are important ramifications with this concept in regard to both the material and sectional properties. They are assumed for members during the analysis, and also during the translation of the results which are derived using elastic modelling techniques into the inelastic response domain.

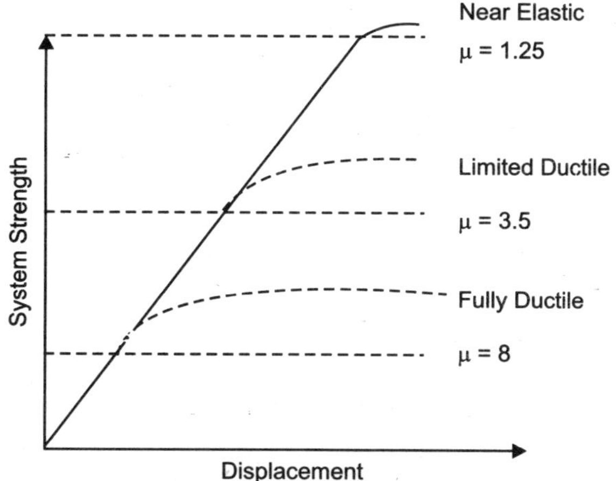

Fig. 11.1 Post-elastic (Ductility) system capacities (Courtesy: http://www.branz.co.nz)

For compliance with the serviceability limit state performance provisions, the simple linear stress/strain relationships of materials are needed. These are the conventional parameters used to assess the structural resistance to other loads. Assuming that the structural system remains predominantly elastic, damage avoidance can reasonably be expected and compliance can be assured. Simple elastic engineering models can be used to ascertain building response in these conditions. Consequently, for concrete and masonry structures, the cracked sectional properties are appropriate for the serviceability limit state, although significant yield of the reinforcing steel (and the subsequent retention of wide residual cracks) is to be avoided.

For compliance with the ultimate limit state performance provisions, the post-elastic response of the structure needs to be considered. This includes large post-elastic member deformation (Fig. 11.1). Often traditional engineering models break down at this stage. There is thus little to be gained by using highly sophisticated engineering modeling techniques to demonstrate compliance with the ultimate limit states criteria unless there is a high degree of confidence that the relationship between the elastic and inelastic structural response is realistic. The simple elastic stress/strain relationships and the elastic engineering models used to ascertain the load distribution between members within the structural system no longer apply. It is to address this particular post-elastic response condition, being the primary objective of good earthquake engineering design, that the principles of capacity design of structures were developed and subsequently introduced.

11.4 EARTHQUAKE DESIGN LEVEL GROUND MOTION

A fundamental parameter contained within all earthquake loading standards is the earthquake induced ground motion. This has to be used for design. This is generally prepared by seismologists and geotechnical engineers. It is typically presented to the structural designer in three components. They are the elastic response of the basement rock (usually as acceleration spectra), the relative seismicity at the site (commonly presented as a suite of zonation maps), and a modification function which is applied to the motion at bedrock beneath the site to allow for near surface soil conditions (presented as either a simple amplification factor or as a more complex soil property related function).

11.4.1 Elastic Response Spectra

Engineers traditionally have used acceleration response spectra to represent the motion induced by the design earthquake. These spectra are generally presented as a response function (acceleration, velocity or displacement) against the response period of a single-degree-of-freedom oscillator (refer Fig. 11.2). Spectra are developed by calculating the response of a single mass oscillator (usually with 5% critical damping present) to the design level earthquake motion. Engineers traditionally have shown a preference for acceleration spectra, since the resulting coefficient, when multiplied by the seismic mass, results in the lateral base shear for the building. In Australia, and the uniformed building code used in the western USA, these spectra are presented as a simple uniform coefficient followed by an exponential decay. The New Zealand Loadings Standard prescribes an elastic response spectrum, derived using a uniform risk approach, for each soil class. The modern trend as indicated by the European

Earthquake Standard and also in the proposed National Earthquake Hazard Reduction Programme (NEHRP) specification is to acknowledge that the response spectra is building period dependent. This is achieved by publishing the design spectra in parametric form. In parametric form, the ordinates of each parameter and the characteristics of the curve between them are read from a series of seismic zonation maps of the region.

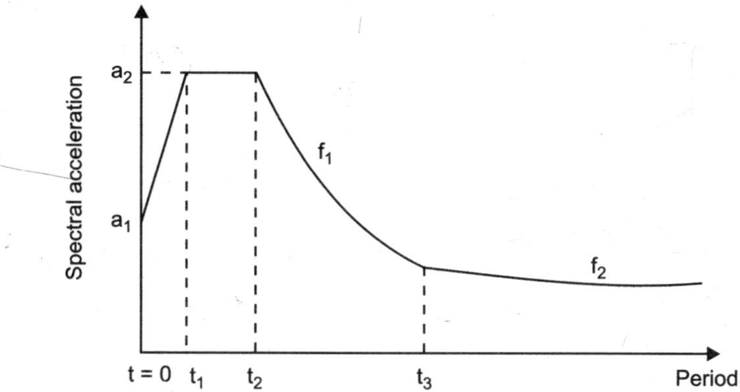

Fig. 11.2 A form of parametric acceleration response spectrum (Courtesy: http://www.branz.co.nz)

11.4.2 Relative Seismicity

The current generation of earthquake loading standards uses a single seismic zonation map. The map has iso-seismal contours to represent the relative seismicity between locations. An example of one for New Zealand is shown in Fig. 11.3. The product of the zone factor, Z, and the lateral acceleration coefficient derived from the design spectrum is used for design. The next generation of earthquake loading standards are expected to specify spectral acceleration as a function of the response period. They will also design event return period. The simple linear scaling of a standard spectral shape will no longer be acceptable. Instead we may expect, for example, a suite of three series of maps to reflect different probabilities of exceedance (0.05 (20 year return period) 0.002 (500 year return period) and 0.0005 (2000 year return period)). Each set will comprise 4 maps each with spectral ordinates for periods of perhaps T = 0, T = 0.2 seconds, T = 1 second and T = 2.5 seconds. The complete suite may therefore comprise 12 regional maps which will enable the development of different shaped elastic response spectra for different return periods.

This approach is likely to have significant impact on regions of low to moderate seismicity. Reference to Fig. 11.4 indicates that while, as expected, the peak ground acceleration is much higher in high seismicity regions than it is in low seismicity ones (ratio of 3.5:1). The differential is markedly reduced as the probability of exceedance increases (2:1 for 0.0005 probability of exceedance) with the PGA being approximately equal to that of the normal design event within a high seismicity area. It is likely that important key facilities of the future will be required to survive earthquakes with exceedance intervals of this order. The design requirements may well be quite similar regardless of regional seismicity in such events.

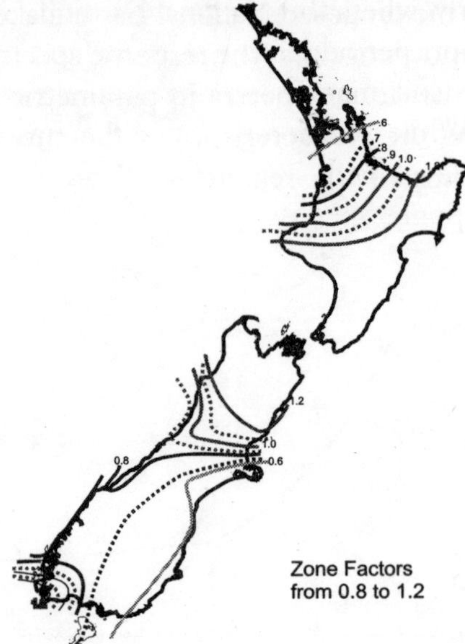

Fig. 11.3 Typical seismic zonation map (interpolation between iso-seismals is acceptable) (Courtesy: http://www.branz.co.nz)

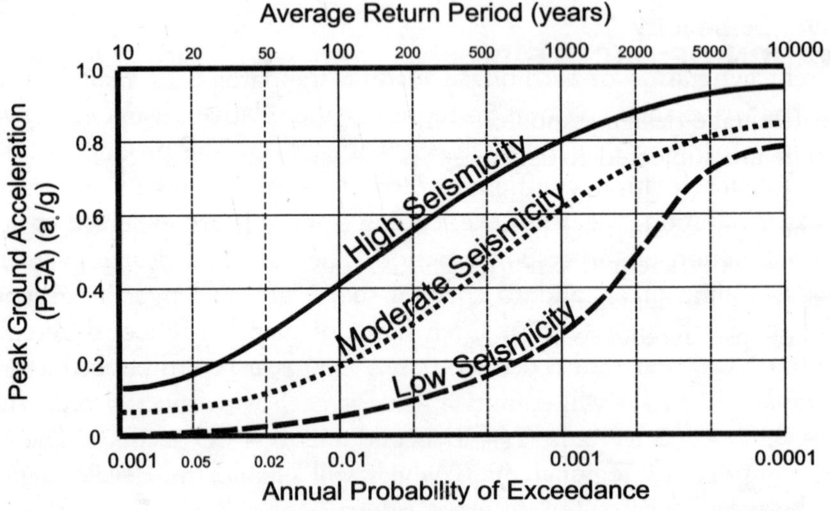

Fig. 11.4 Variations of PGA against probability of exceedance with seismicity (Courtesy: http://www.branz.co.nz)

11.4.3 Soil Amplification

Earthquakes are usually initiated by rupture over a fault rupture plane. They are often deep within the earth's mantle. The ground motion experienced on the surface results from the transmission of energy waves released from that bedrock source. It is transmitted first through bedrock and then undergoes significant modification by soil layers as the energy waves near the earth's surface. Typically rock sites experience high short period response but more rapid decay. Thus, short duration high intensity motion may be expected in such locations.

Conversely soft soils, particularly when they extend to moderate depths (>50 metres) are likely to filter out some of the short period motion. Usually longer period response is amplified, particularly in cases where the soil mass has a natural period similar to the high energy component of the earthquake. While such resonance effects can be taken into account when site specific spectra are being developed, it is usually impractical to include such effects in a loading standard. Soft soil response spectra have a flatter, broader plateau (refer Fig. 11.5).

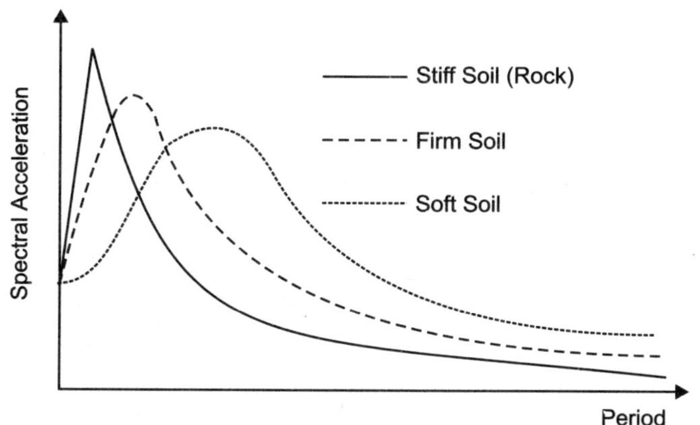

Fig. 11.5 Typical spectral response curves modified for soil effects (Courtesy: http://www.branz.co.nz)

11.5 DERIVATION OF DUCTILE DESIGN RESPONSE SPECTRA

Most modern earthquake design standards acknowledge the fact that buildings will experience damage when they are subjected to severe earthquake attack. Attempts are made to quantify the post-elastic capacity of different building type as well as material types. This is achieved by including some form of ductility based adjustment factor. This has the effect of reducing the elastic response coefficient down to a more convenient level. Below this level, elastic response with little or no damage is expected. However, beyond this level, some damage is accepted while collapse avoidance is to be assured.

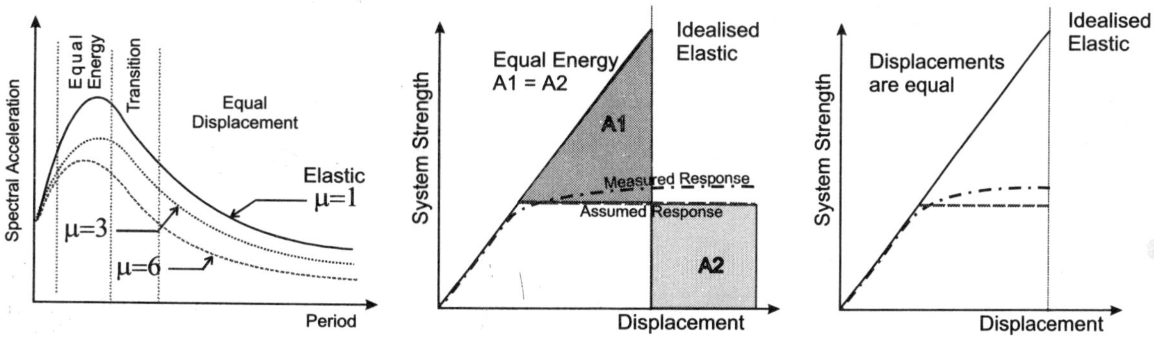

Fig. 11.6 Basis for translation of elastic response spectra to inelastic design spectra (Courtesy: http://www.branz.co.nz)

Earthquake standards differ in how they translate the elastic response spectra derived for the site into inelastic spectra which can be used as the basis for structural design. This includes both the seismic zonation factor and the local soil factors The two most common methods are to use a combination of structural ductility and structural performance factors. Within the New Zealand Loadings Standard, this is a combination of the ductility factor, μ, and the structural performance factor, Sp. The European Earthquake standard, combines these as a structural behaviour factor, q. The earthquake standards of Australia and the UBC used in the western USA use a structural response factor, R_f. Both q and R_f are period independent. Consequently, they are direct scaling factors of the site response spectra. The various inelastic response spectra published within the New Zealand Standard introduce period dependency with equal energy concepts being applied to short period structures as well as equal displacement to long period ones. Furthermore, there is a transition zone in between (refer Fig. 11.6). For very long period structures, a constant displacement response can be expected.

11.6 ANALYSIS AND EARTHQUAKE RESISTANT DESIGN PRINCIPLES

11.6.1 The Basic Principles of Earthquake Resistant Design

Earthquake forces are generated by the dynamic response of the building to earthquake induced ground motion. This makes earthquake actions fundamentally different from any other imposed loads. Thus the earthquake forces imposed are directly influenced by the dynamic inelastic characteristics of the structure itself. While this is a complication, it provides an opportunity for the designer to heavily influence the earthquake forces imposed on the building. Through the careful selection of appropriate, well distributed lateral load resisting systems, the influence of many second order effects, such as torsional effects, can be minimised. Furthermore, by ensuring the building is reasonably regular in both plan and elevation, significant simplifications can be made to model the dynamic building response.

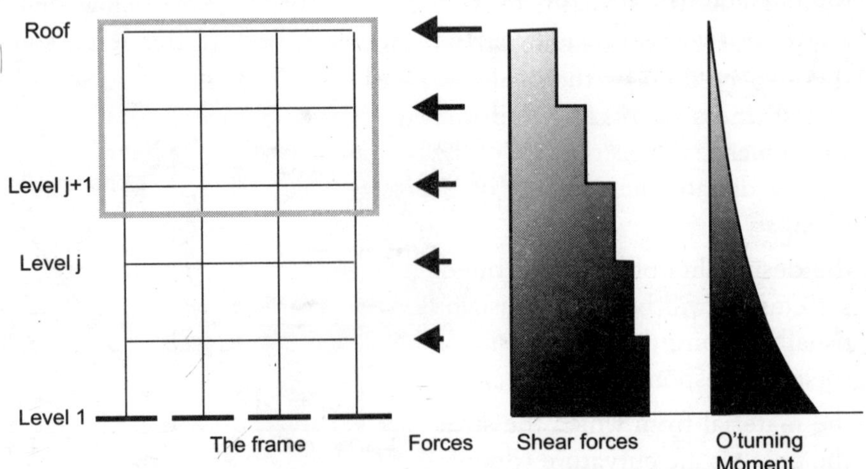

Fig. 11.7 Loading pattern and resulting internal structural actions (Courtesy: http://www.branz.co.nz)

Most buildings can be reasonably considered as behaving as a laterally loaded vertical cantilever. The inertia generated earthquake forces are generally considered to act as lumped masses at each floor (or level). The magnitudes of these earthquake forces are usually assessed as being the product of seismic mass (dead load plus long-term live load) present at each level and the seismic acceleration generated at that level. The design process involves ensuring the resistance provided at each level is sufficient to reliably sustain the sum of the lateral shear forces generated above that level (refer Fig. 11.7).

11.6.2 Controls of the Analysis Procedure

A schematic of the earthquake design process is presented in Fig. 11.8. The essential features of the process are as follows:

1. Structural designers are usually given the site location. They are also provided with intended occupancy of the building.
2. The national building code normally includes the requirements for the following:
   (i) the design philosophy acceptable for buildings (Limit States or Working Stress Design)
   (ii) the performance objectives for the prescribed occupancy class.
   (iii) the structural importance classification and
   (iv) the proportion of live load considered to be present during a major earthquake.
3. Derive the peak ground acceleration (i.e. elastic response spectrum for T=0) for the design intensity earthquake ground motion. The derivation is from the consideration of the seismicity of the region, modified by the near surface soil modification factor (refer Section 11.4). The seismicity of the region is selected to match the design event return period
4. Select a suitable structural configuration. This selection is with consideration for the following parameters:
   (i) the characteristics of the various lateral load resisting structural forms available.
   (ii) the desirability of matching the strength and stiffness of the structural frame to that expected under the dynamic loading of the building itself. It implies strength and stiffness decreasing uniformly up the height of the building. However, this will influence the distribution of the base shear over the building height. Consequently, it may dictate the method of analysis acceptable for the building to avoid its collapse.
   (iii) the desirability of a regular building plan with well balanced lateral load resisting systems. It will be evenly distributed about the building plan. Irregular plan will usually require three dimensional analysis. Furthermore, it may experience severe torsional response.
   (iv) the material from which the structural system is to be constructed. Consequently, the post-elastic curvature (ductility) which can be accommodated through specific detailing.
5. Determine the level of design required. (Note: There will be many normal occupancy buildings in regions of low seismicity. They do not require any specific earthquake resistant measures to be introduced. Other levels of design involve a) simply tying

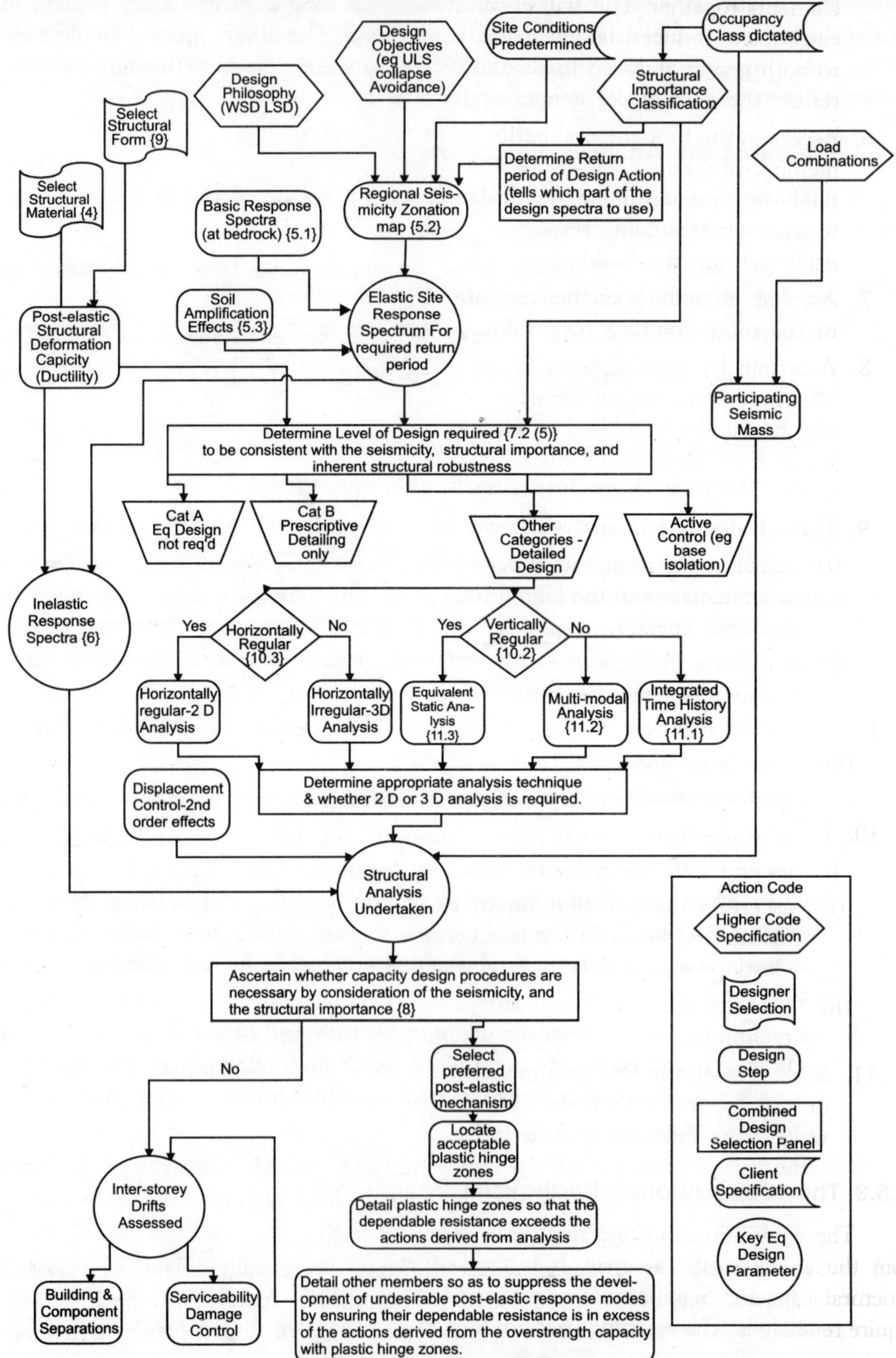

Fig. 11.8 Schematic of the earthquake design procedure (Courtesy: http://www.branz.co.nz)

elements together. This will ensure a continuous, rational load path which exists for earthquake induced lateral loads, or, b) detailed analysis of the building subjected to both gravity induced loads and a rationally derived lateral loading pattern. It will reflect the earthquake generated forces.

6. Ascertain the fundamental period of response of the building. It is based on assumed member of sections and properties. (Note: Several empirical formula are available as the basis for determining the fundamental period of buildings. It is generally preferable to assess the building response based on a realistic distribution of seismic mass at each level up the building).
7. Ascertain from the horizontal regularity of the structure, whether a simple two dimensional or the more complex three dimensional analysis model required.
8. Ascertain by consideration of the vertical regularity of the structure, whether the structural response will be dominated by the first mode response of the structure in which case the simplified equivalent static design procedure can be used. Otherwise, because of vertical structural irregularity, multi-modal analysis will be required to enable the base shear distribution to be established.
9. If equivalent static analysis is acceptable then:
   (i) calculate the design level base shear force. It is obtained from the product of the seismic mass and the lateral force coefficient which are derived from the inelastic response spectra.
   (ii) distribute the base shear to each level of the building and between lateral load resisting systems. This is done in accordance with horizontal and vertical regularity of the structure.
   (iii) use elastic analysis techniques to determine actions induced on members from load combinations which include earthquake forces.
10. If multi-modal analysis is required then:
   (i) ascertain the period and deformed shape for each mode.
   (ii) ascertain the contribution of each mode from the base shear of each mode, distributed between levels according to each mode shape. It is derived from the elastic response spectra lateral acceleration at each respective modal period.
   (iii) combine the contribution from each mode using an appropriate modal combination technique.
11. Scale the elastic deformation obtained from the analysis to allow for post-elastic deformations. Furthermore, check that the overall deformation of the structure as well as the inter-storey drift limits are within acceptable limits.

11.6.3 The 'Conventional' Earthquake Design Procedure

The conventional engineering design approach is to use the actions for members derived from the above elastic analysis. It is used as the basis for determining the dimensions and structural capacity. Significant changes in dimension will affect the building stiffness. It may require re-analysis. The resulting sizes are then checked against those assumed during the analysis.

Earthquake design has three important distinctions from other loadings. Firstly there is the acceptance that damage to both non-structural and some structural elements will occur. However, collapse is to be avoided. Secondly, earthquakes are highly variable dynamic events which designers tend to simplify into a set of quasi-static lateral loads. This approach enables relatively simple analysis and design. However, it noticeably departs from reality. It is therefore important to build into the structure a degree of toughness or robustness which will avoid the development of undesirable collapse mechanisms. Thirdly, although there is geological and seismological understanding of how earthquakes are initiated and how the energy release mechanisms translate into surface ground motion, earthquakes still inherently contain a higher level of uncertainty.

Several modern earthquake design standards, particularly those which apply to regions of moderate or high seismicity, permit designers to make special provisions to accommodate the anticipated level of damage. They also take additional measures to ensure the collapse prevention mechanisms are robust enough to avoid overloading. This can be achieved when the designer takes control of the structure, and dictates which and where post-elastic mechanisms are to occur. The designer should also ensure the post-elastic demands are within levels acceptable for the material being used. Furthermore, undesirable possible collapse mechanisms are to be suppressed within the elements themselves.

11.7 EARTHQUAKE RESISTANT STRUCTURAL SYSTEMS

Three types of earthquake resistant structural systems are generally available.

11.7.1 Moment Resisting Frames

Moment resisting frames typically comprise floor diaphragms supported on beams which link to continuous columns. The joints between beam and columns are usually considered to be 'rigid'. The frames are expected to carry the gravity loads through the flexural action of the beams as well as the propping action of the columns. Lateral loads, imposed within the plane of the frame, are resisted through the development of bending moments in the beams and columns. Framed buildings often employ moment resistant frames in two orthogonal directions. The column elements are common to both frames in such cases.

Moment resisting frames are well suited to accommodate high levels of inelastic deformation. When a capacity design approach is employed, it is usual to assign the end zones of the flexural beams to accept the post-elastic deformation expected. Furthermore, the column members are designed such that their dependable strength is in excess of the over-strength capacity of the beam hinges. This ensures that they remain within their elastic response range regardless of the intensity of ground shaking due to earthquake. Moment resisting frames are, however, often quite flexible. When they are designed to be fully ductile, special provisions are often needed to prevent the premature onset of damage to non-structural components of the structure.

11.7.2 Shear Walls

The primary function of shear walls is to resist lateral loads. However, they are often used in conjunction with gravity frames and carry a proportion of gravity loads also. Shear walls fulfil their lateral load resisting function by vertical cantilever action. By reference to Fig. 11.3 it can be seen that both the shear force and bending moment generated by the earthquake actions increase down the height of the building. Since shear walls are generally both stiff and can be inherently robust, it is practical to design them to remain nominally elastic under design intensity loadings. This is particularly true in regions of low or moderate seismicity. Under increased loading intensities, post-elastic deformations will develop within the lower portion of the wall. This portion is generally considered to extend over a height of twice the wall length above the foundation support system. However, it can result in difficulties in the provision of adequate foundation system tie-down to prevent uplift. Consequently, the design of rocking foundations is common with shear walls. Good post-elastic response can be readily achieved within the region of reinforced concrete or masonry shear walls. It is achieved through the provision of adequate confinement of the principal reinforcing steel as well as the prohibition of lap splices of reinforcing bars.

Shear wall structures are generally quite stiff. Inter-storey drift problems are rare and generally easily contained. The shear wall tends to act as a rigid body rotating about a plastic hinge. The hinge forms at the base of the wall. Overall structural deformation is thus a function of the wall rotation. Inter-storey drift problems which do occur are limited to the lower few floors. A major shortcoming with shear walls within buildings is that their size provides internal (or external) access barriers. It may contravene the architectural requirements. This problem can be alleviated by coupling adjacent more slender shear walls. The coupling beams then become shear links between the two walls. Their careful detailing can provide a very effective, ductile control mechanism.

11.7.3 Braced Frames

Frames employing diagonal braces as the means of transmitting lateral load are common in low-rise and industrial buildings. The bracing elements are typically inclined axially loaded members which traverse diagonally between floors and column lines. They are very efficient in direct tension. Furthermore, they may also be detailed to accept axial compression. However, suppression of compression buckling requires careful assessment of element slenderness. Major shortcomings of braced systems are that their inclined diagonal orientation often conflicts with conventional occupancy use patterns, either internally, across windows or external fabric penetrations. Furthermore, they often require careful detailing to avoid large local torsional eccentricities being introduced at the connections with the diagonal brace being offset from the frame node. A variation on this form of lateral resisting system is the eccentrically braced frame. This system employs a horizontal 'K' form of bracing. The central zone of the 'K' acts in flexure as the tension/compression legs of the brace drive the beam element into direct flexure.

11.8 THE IMPORTANCE AND IMPLICATIONS OF STRUCTURAL REGULARITY

Most Standards outline certain provisions relating to the vertical regularity of the structure, as well as the plan regularity. These usually apply to the appropriateness of several assumptions. These assumptions are implicit in the distribution pattern of the loading as well as the torsional effects.

11.8.1 Vertical Regularity

Ideally the capacity of the structure should follow the shear and bending moment pattern of the structure shown in Fig. 11.3. Substantial departures from this ideal typically result in the onset of premature post-elastic deformations. They are often concentrated at over one level. When this occurs, elements within the one level degrade. It attracts additional post-elastic deformation. Consequently, a soft-storey mechanism develops with collapse often being the inevitable result.

The vertical regularity check is intended to avoid abrupt changes in overall strength as well as the stiffness at any particular level. Where such provisions are not met, more detailed analysis will be required. Objective of such analysis is to ensure that post-elastic deformation capacity at each level can be met without unacceptable loss of strength. Furthermore, post-elastic deformation demands in excess of their capacity should also be met.

It is wise to avoid abrupt curtailment of reinforcing steel at one level of a reinforced concrete frame or substantive changes in a column section. It is better to introduce such changes gradually, over several floors. This allows a smooth transition between sections to develop. Obviously it is undesirable to curtail shear walls above their base, since it induces a very real potential for soft-storey development.

11.8.2 Horizontal Regularity

The random, three dimensional motions generated by earthquakes is usually simplified into two transverse orthogonal components. The vertical response is typically ignored. The transverse dynamic response may also introduce twisting and torsional effects. This could either directly be as a function of the input ground motion, because of variations in the spatial distribution of seismic mass, or because of the structure being irregular in plan.

Measures employed to counter these effects typically involve distributing the lateral load resisting systems about the building plan. Furthermore, attempt is made to limit the plan profile to being reasonably regular and compact. Most modern earthquake loading standards require the designer to assess the Centre of Rigidity (CoR) of the structural system. The centre of mass (CoM) of the uniformly distributed seismic mass should also be assessed. The eccentricity is typically increased by 10% of the building width. This is done to allow for unexpected variations in torsional effects with the magnitude of the resulting torsional action increasing accordingly. The torsional action is the product of mass and linear eccentricity between CoR and CoM. Such approximations tend to be based upon the response of the structure within the elastic response domain. However, they may provide little security against collapse once the deformations have progressed into the inelastic domain. (Paulay, 1997), has recently proposed an elegant means of directly addressing post-elastic torsional effects. He postulates that the post elastic torsional demand can be met, satisfied and indeed controlled by rigorous detailing of lateral load resisting elements. This ensures that their displacement ductility demands are met. Provided this is achieved, the effects of torsion are readily accommodated.

The preferred method of minimising torsional effects is to select floor plans which are regular and reasonably compact. Wide separation of horizontal lateral load resisting systems is encouraged. Plan forms with re-entrant corners such as 'L' and 'T' plan layouts should be avoided. If these plan forms are dictated by other constraints, seismic separation joints should be introduced between rectangular blocks. Such joints must be designed to accept the post-elastic dynamic response of the building parts. This may be responding with disparate phases. Contact and hammering between blocks is also to be avoided.

11.8.3 Floor Diaphragms

The floor system, in additional to supporting the live loads induced by the building contents, can also be designed as a floor diaphragm. This floor diaphragm links the lateral load resisting systems at each level. Other horizontal loading mechanisms such as horizontal trusses or deep beams can be used where the floor system is interrupted by penetrations or openings. The diaphragm action of floors is often taken for granted during design. It is important that designers allow for the concentrations of horizontal stress within the floor diaphragm around openings. Care should also be taken to ensure that the interconnection between the floor diaphragm and the vertical lateral load resisting system is sufficiently robust. This helps to transmit the required shear between elements. Precast flooring systems using slender cast in situ toppings can be quite vulnerable in these conditions.

11.9 METHODS OF ANALYSIS

Earthquake engineering design techniques have advanced greatly. This has occurred with the advent of modern computing techniques. The prudent designer, however, is wise not to lose sight of the primary objective of earthquake design (i.e. collapse avoidance). Furthermore, the level of uncertainty present in several of the key input parameters should also be taken into account. Little may therefore be achieved by using highly sophisticated analysis techniques. It has been concluded that the selection of regular building configurations and the application of sound detailing principles are more likely to provide the required level of security against collapse. On the other hand, detailed refinement of the analysis techniques will not be useful.

11.9.1 Integrated Time History Analysis

Integrated time history analysis techniques involve the stepwise solution in the time domain of the multi-degree-of-freedom equations of motion. These equations represent the actual response of a building. It is the most sophisticated level of analysis available to the earthquake engineer. Its solution is a direct function of the earthquake ground motion, which are selected as the input parameter for the specific building. Such records are seldom available directly for a given site. On the other hand, either synthetic ground motion or modified real free-field records are generally used. The modelled representation of the structure itself is required to realistically represent both the elastic and post-elastic response characteristics of the building. However, this detail information is seldom available when commencing the design process. Consequently, this analysis technique is usually limited in its application to checking the suitability of assumptions made during the design of important structures.

11.9.2 Multi-modal Analysis

Multi-modal analysis is an elastic dynamic response analysis technique. This involves first the determination of the structural response of each mode of vibration of the building. Then it is followed by the combination of the resulting forces for each significant response mode.

For such assessments it is usually convenient to consider the structural mass to be concentrated at each floor level. This results in one degree of freedom for each floor provided torsional effects are ignored. When the building is torsionally susceptible, lateral and torsional response will need to be considered. Consequently, the number of possible response modes will be doubled. The procedure involves determining both the response period and the mode shape. The determination of the lateral shear coefficient for each response mode (from the design spectra using the modal period) and the distribution of the resulting base shear according to the response shape at each floor is also involved. The contribution of each response mode is then combined with an allowance being made for the time variance between different response modes. Thus a square-root-sum-of-squares (SRSS) method of combining lateral forces is generally used. Other combination methods, such as the complete quadratic combination (CQC) method may be required where the response modes are close together. A static analysis using the resulting equivalent forces is also used as the basis for determining the forces and displacements of the overall structure.

This technique takes into account allowance for the true response characteristics of the building. However, it should be remembered that it is still assessing the structural response while it remains elastic. Collapse avoidance, with the implied onset of controlled damage (i.e. post-elastic deformations), requires many assumptions to be made to arrive at the inelastic response. In addition many of the structural member properties needed for the analysis are unknown. They can be known only after a preliminary analysis has been undertaken. Consequently, the sophistication of the model often adds little to the final design.

11.9.3 Equivalent Static Analysis

The equivalent static analysis procedure is also essentially an elastic design technique. However, some consideration of the post-elastic response enters into the selection of the determination of the lateral force coefficient. It is, however, simple to apply the multi-model response method. The implicit simplifying assumptions are arguably more consistent with other assumptions implicit elsewhere in the design procedure. The equivalent static analysis procedure involves the following steps:

1. Estimate the first mode response period of the building. This is obtained from the design spectra.
2. Use the specific design response spectra to determine the lateral base shear of the complete building. This should be consistent with the level of post-elastic (ductility) response assumed.
3. Distribute the base shear between the various lumped mass levels. It is usually based on an inverted triangular shear distribution of 90% of the base shear. 10% of the base shear is being imposed at the top level to allow for higher mode effects.

4. Analyse the resulting structure under the assumed distribution of lateral forces. Furthermore, determine the member actions and loads.
5. Determine the overall structural response. Particularly regarding the inter-storey drifts assessed for the elastically responding structure. For the assessment of the post-elastic deformation, design standards typically magnify the elastic deformed shape by the structural ductility. This helps to determine the overall maximum deformation. Usually it is at the roof level. Introduction of a non-linear response profile to allow for local rotation at plastic hinge zones is often required when determining the inter-storey drifts.

Home Work Problems

1. Comment on key material parameters for effective earthquake resistant design for compliance with:

   (*i*) serviceability limit state

   (*ii*) ultimate limit state
2. Explain about components of earthquake design level ground motion.
3. Write short note on ductile design response spectra.
4. What are the types of earthquake resistant structural systems? Briefly explain about them.
5. Comment on earthquake engineering design techniques.

12 CHAPTER

EARTHQUAKE-INDUCED SETTLEMENTS IN SATURATED SANDY SOILS

12.1 INTRODUCTION

In this chapter, a simplified approach is proposed to estimate the earthquake-induced settlements in saturated sandy soils. The simplified approach consists of a set of equations. These equations can be used to estimate directly the settlements of a saturated sandy soil layer subjected to earthquake loading. The proposed approach is based on Tokimatsu and Seed (1987) procedure, but does not require the use of charts, tables and diagrams. Consequently, it can be conveniently implemented numerically and allows the expeditious analysis of complicated soil profiles. It is particularly useful for sensitivity analyses as well as for analyzing soil profile with multiple layers. The proposed approach may be able to characterize the liquefaction boundary line. This boundary line separates liquefaction from non-liquefaction regions. The proposed approach has been used to predict the settlements observed at two sites which were subjected to different magnitude of earthquakes. The settlement predictions by the proposed method are in reasonably good agreement with those computed by other published approaches as well as with field measurements.

The tendency of sands to densify when subjected to earthquake loading is well established. The post-earthquake densification of saturated sand is influenced by the grain size and relative density of the sand, the maximum shear strain induced in the sand, and the amount of excess pore pressure generated by the earthquake. The dissipation of excess pore pressure induces consolidation settlement of the ground (Lee and Albaisa 1974).

A number of procedures have been presented in the literature in the past 20 years to study the earthquake-induced settlement problem. They can vary from relatively complex non-linear dynamic computer models (e.g. DESRA or TARA3FL) to simplified procedures. The simplified procedure estimates the consolidation settlement from volumetric strain based on the cyclic stress ratio and normalized SPT-N value. Tokimatsu and Seed (1987) reviewed available procedures and recognized that the primary factors controlling earthquake-induced settlement are the cyclic stress ratio with pore pressure generation, together with the corrected SPT-N value, and earthquake magnitude. Using a correlation between the normalized SPT-N value and relative density as well as an estimate of the shear strain potential of liquefied soil from the normalized SPT-N and cyclic stress ratio (Seed *et. al*, 1984), Tokimatsu and Seed (1987) developed a chart that allows the volumetric strain after liquefaction in a magnitude of 7.5 earthquake to be estimated directly from the cyclic stress ratio and normalized SPT-N value. Alternatively, Ishihara and Yoshimine (1992) developed a graphic representation when estimations are required at sites having multiple layers of soil with different properties.

12.2 METHOD OF ANALYSIS

12.2.1 Cyclic Stress Ratio (CSR)

The cyclic stress ratio, CSR, is given by (Seed et al., 1985):

$$CSR_{7.5} = 0.65\left(\frac{\sigma_v}{\sigma_v'}\right)\frac{(a_{max})}{g}(r_d)/MSF \qquad ...(12.1)$$

Where,

$CSR_{7.5}$ = cyclic stress ratio with reference to earthquake magnitude of 7.5

σ_v, = total overburden pressure at the depth considered

σ_v' = effective overburden pressure at the same depth

a_{max} = maximum horizontal acceleration at the ground surface

g = acceleration due to earth's gravity

r_d = stress reduction factor, and

MSF = magnitude scaling factor

12.2.2 Volumetric Strain After Liquefaction

Excess pore water pressure may be build up when saturated sandy ground is subjected to a sequence of cyclic shear stresses. These cyclic shear stresses are induced by earthquake loading. The dissipation of this excess pore pressure results in reconsolidation volumetric strain (Lee and Albaisa, 1974). The amount of reconsolidation volumetric strain increases with decreasing density of soils and increases with increasing maximum shear strain developed in the deposit. However, it is independent of vertical effective stress (Tatsuoka *et al.*, 1984).

Among the various field tests, Standard Penetration Test (SPT) is commonly preferred because of the difficulty and expense associated with obtaining undisturbed field samples of sandy soils. Furthermore, the more extensive databases of sites which have experienced obvious liquefaction, as well as those where no apparent liquefaction occurred have been evaluated for a number of seismic events. They have been related to the Standard Penetration Tests (SPT). The criteria for evaluation of liquefaction behaviour based on SPT appear to be rather robust over the years.

Tokimatsu and Seed (1987) reviewed available procedures and developed an empirical chart that allows the estimate of earthquake-induced volumetric strains in saturated sands based on cyclic stress ratio and normalized SPT values. Based on their findings, the volumetric strains due to soil liquefaction during an earthquake magnitude of 7.5 may be approximated by the equation:

$$\varepsilon_v = 10[(N_1)_{60}]^{-0.6} \text{ for } \left(\frac{CSR}{(N_1)_{60}}\right) > 0.01 \qquad \text{...(12.2)}$$

Where, ε_v = volumetric strain and $(N_1)_{60}$ = SPT–N value normalized to an effective overburden of 1 tsf (95.76 kPa) and to an effective energy of 60% of free-fall energy.

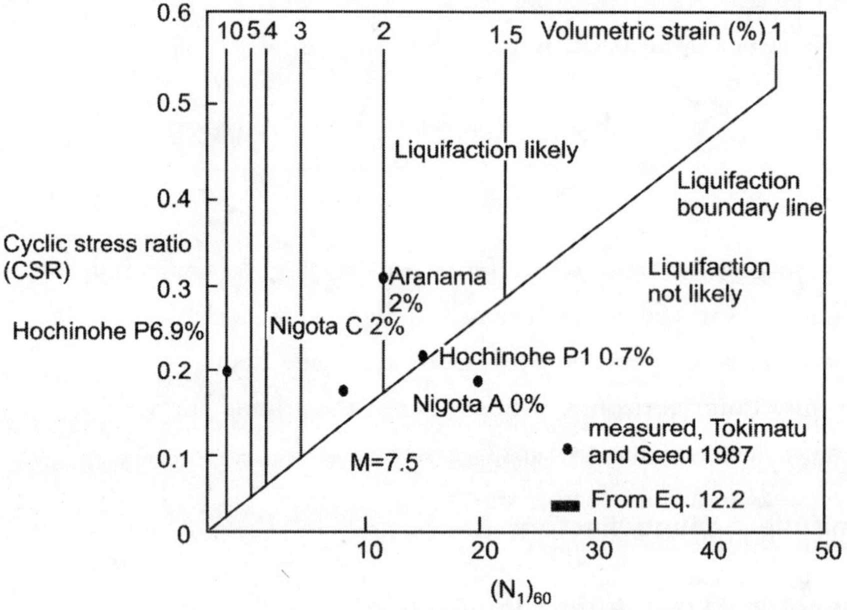

Fig. 12.1 Relation between cyclic stress ratio, $(N_1)_{60}$ & volumetric strain for saturated sands

(Courtesy: http://www.arpnjournals.com)

Figure 12.1 illustrates a liquefaction boundary line as well as a set of simplified volumetric strain lines defined by equation (12.2). Also plotted and presented in Figure 12.1 are measured results from case history information. The computed volumetric strain lines appear to be generally consistent with the measured field cases. It is observed that the data points falling below the liquefaction boundary line are not likely to experience liquefaction behaviour. Hence it may provide a simple way to explore the level of risk of liquefaction based on the cyclic stress and SPT values. However, this observation may or may not be valid in cases other than those in the data set analyzed. This is because the simplified volumetric strain lines are drawn based on limited data points.

Figure 12.2 shows the volumetric strains (ε_v) versus $(N_1)_{60}$ computed by equation (12.2). The computed values agree reasonably well with those suggested by Tokimatsu and Seed (1987). However, it is evident from Figure 12.2 that equation (12.2) appears to compute higher volumetric strains than those suggested by Tokimatsu and Seed (1987) at $(N_1)_{60}$ value in excess of 30. Nevertheless, the magnitude of the volumetric strain occurring at $(N_1)_{60}$ greater than 30 may be insignificant.

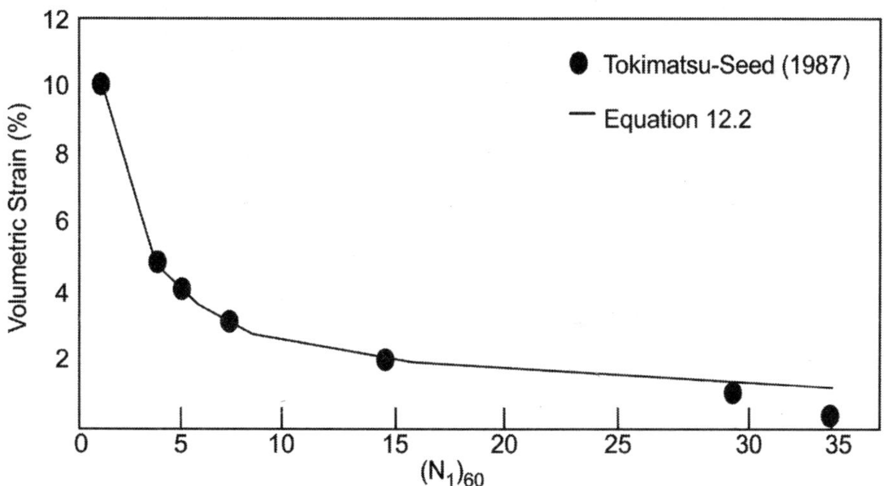

Fig. 12.2 Relationship between volumetric strain and $(N_1)_{60}$
(Courtesy: http://www.arpnjournals.com)

12.2.3 Magnitude Scaling Factors (MSF)

The magnitude scaling factors (MSF) have been revised by some researchers (e.g. Arango 1996, Andrus and Stokoe 1997, Youd and Noble 1997, and Idriss 1999). However, the scaling factors suggested by Tokimatsu and Seed (1987) for saturated sands have been employed in present discussion for the comparison with procedures available in the literature which used the similar scaling factors.

The magnitude scaling factors (MSF) based on the data obtained by Tokimatsu and Seed (1987) may be approximated by the equation:

$$MSF = 2.5 - 0.2M \qquad ...(12.3)$$

Where, M = magnitude of earthquake in equation (12.3). The magnitude scaling factors (MSF) computed by using equation (12.3) and those tabulated by Tokimatsu and Seed (1987) are in reasonable good agreement as shown in Figure 12.3.

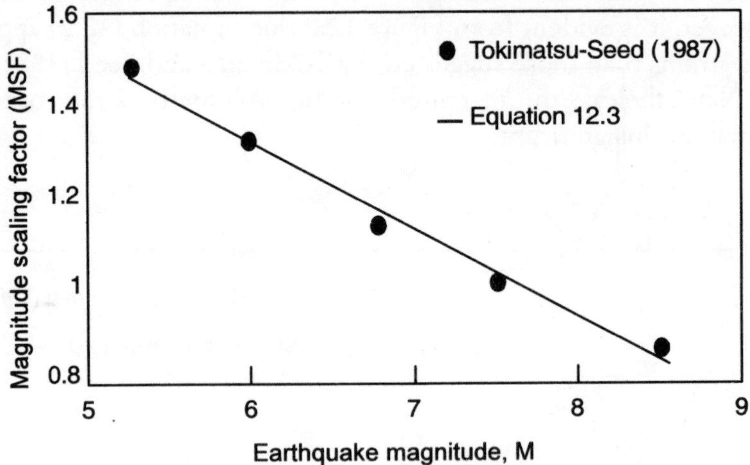

Fig. 12.3 Relationship between magnitude scaling factor and M (Courtesy: http://www.arpnjournals.com)

The total settlement, S, for sandy soil is given by:

$$S = \sum_{i=1}^{n} H_i \varepsilon_{vi} \qquad ...(12.4)$$

Where, H_i = thickness for layer i,

ε_{vi} = volumetric strain for layer i, and

n = number of soil layers.

12.3 COMPARISON OF RESULTS

12.3.1 Tokachi-Oki Earthquake-Site P6 Hachinohe, Japan (1968)

The site P6 at Hachinohe in Japan consists of saturated sand deposit which is extremely loose to a depth of about 6m. The detailed description of the subsurface conditions is described by Ohsaki (1970). The maximum acceleration of 0.2g was measured in vicinity of the site P6 site and M = 7.9. Figure 12.4 shows the typical $(N_1)_{60}$ distribution with depth for the site P6 at Hachinohe (Ohsaki, 1970 and Tokimatsu and Seed, 1987).

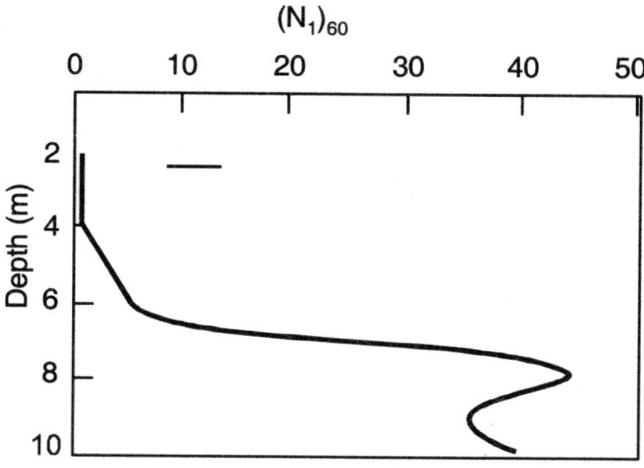

Fig. 12.4 Distribution of $(N_1)_{60}$ with depth
(Courtesy: http://www.arpnjournals.com)

Figure 12.5 illustrates the computed cyclic stress (CSR) ratio versus depth using equation (12.1). The computed cyclic stress ratio (CSR) tends to increase with depth to a depth of about 7m. Beyond 7m depth, it has almost a constant value of 0.225 for larger depths.

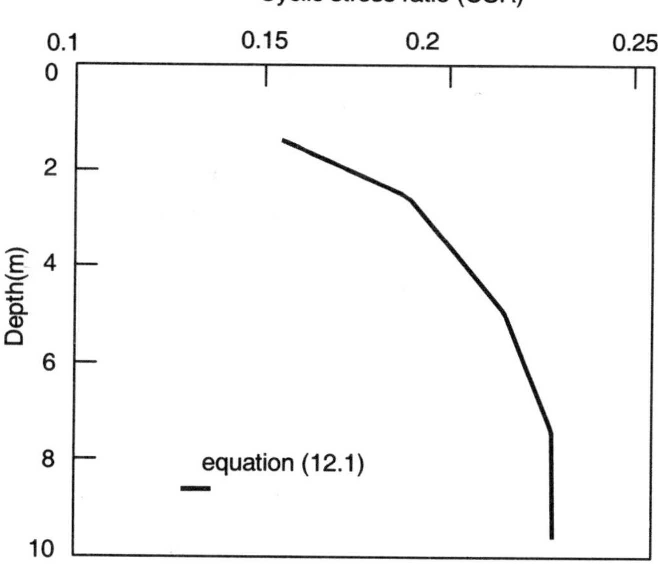

Fig. 12.5 Computed distribution of CSR with depth
(Courtesy: http://www.arpnjournals.com)

The volumetric strains predicted by Tokimatsu-Seed and the approaches discussed in present chapter are shown in (Figure 12.6). The approach discussed in present chapter computes slightly higher volumetric strain than Tokimatsu-Seed approach at the depths of 2–4m. Both approaches predict similar volumetric strains at larger depths.

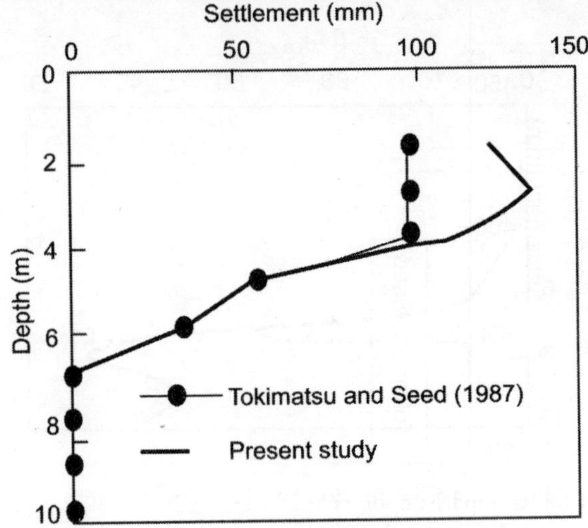

Fig. 12.6 Comparison of computed distributions of volumetric strain with depth (Courtesy: http://www.arpnjournals.com)

Figure 12.7 shows the settlement variation with depth curves predicted by Tokimatsu-Seed as well as by technique discussed in present chapter. As larger volumetric strains are predicted by the present approach at 2-4m depths, it is expected that the corresponding settlements computed by the present approach will also be larger than those computed by Tokimatsu-Seed approach. However at depth greater than 4m both approaches predict similar settlements.

The total settlements predicted by Tokimatsu-Seed as well as by the technique described in present chapter are compared with the measured post-earthquake settlement at this site as shown in Fig. 12.8. It appears that Tokimatsu-Seed approach [Tokimatsu and Seed (1987)] tends to underestimate the settlement by about 11%. Furthermore, the predicted settlement by the present study is closer to the measured settlement. However, both the approaches provide an acceptable design prediction.

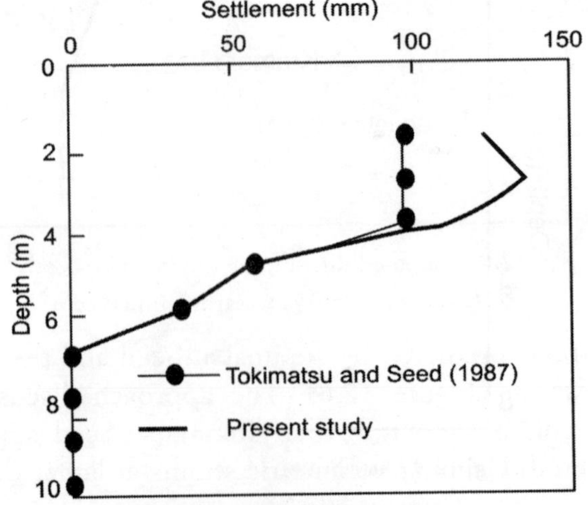

Fig. 12.7 Computed distributions of settlement with depth (Courtesy: http://www.arpnjournals.com)

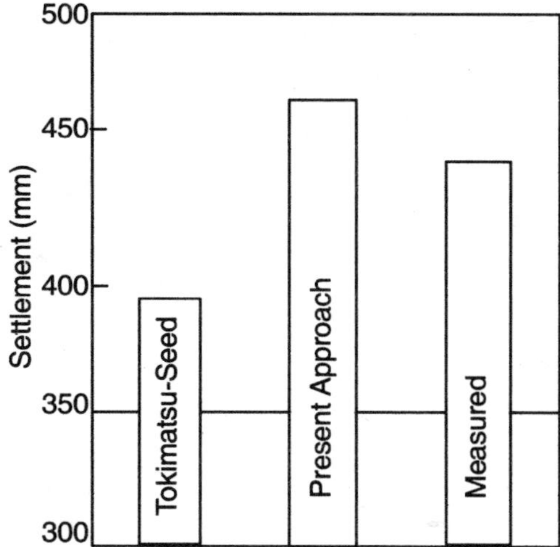

Fig. 12.8 Comparison of computed and measured settlements (Courtesy: http://www.arpnjournals.com)

12.3.2 Loma Prieta Earthquake, San Francisco (1989)

During the 1989 Loma Prieta earthquake, significant ground settlement was observed in the Marina District of San Francisco. It was found that most of this settlement resulted from densification of hydraulic fills. These fills were placed to reclaim the area from San Francisco Bay in the 1890s. The site consisted of about 2m of top fill overlay by 6m of loose sandy fill. A layer of recent bay mud extended from depth 8m to 30m. The water table was located at about 2m below the ground surface. A maximum acceleration of 0.2g was measured in the vicinity of the Marina District. Figure 12.9 shows the $(N_1)_{60}$ distribution with depth for the Marina District site (Kramer, 1996).

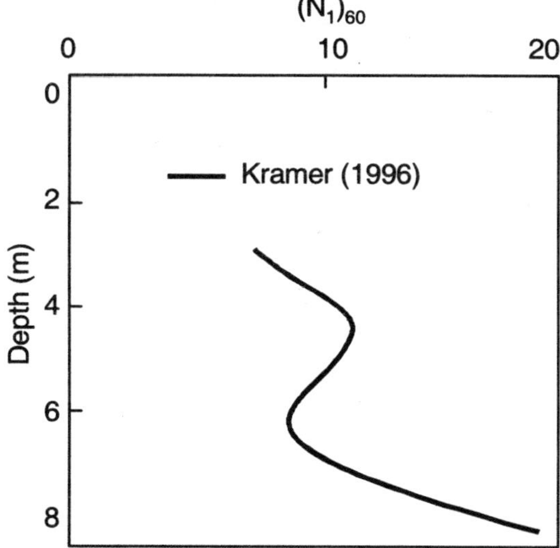

Fig. 12.9 Distribution of $(N_1)_{60}$ with depth (Courtesy: http://www.arpnjournals.com)

Figure 12.10 illustrates the computed cyclic stress ratio (CSR) versus depth using equation (12.1). The computed cyclic stress ratio (CSR) tends to increase with depth.

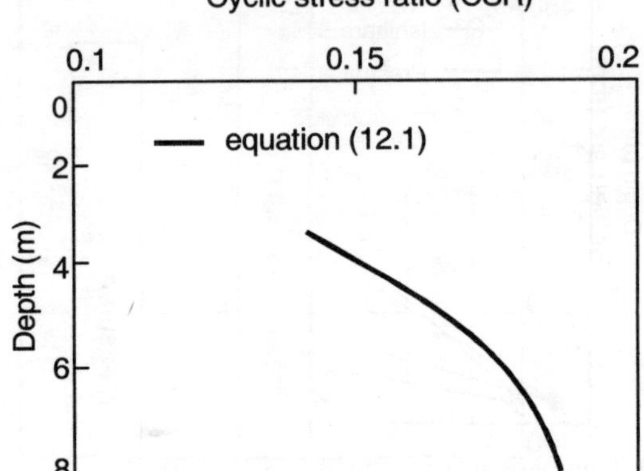

Fig. 12.10 Computed distribution of CSR with depth (Courtesy: http://www.arpnjournals.com)

The volumetric strain predicted by Tokimatsu-Seed, Ishihara-Yoshimine (1992) and the technique described in present study are shown in Figure 12.11. The Ishihara-Yoshimine approach computes higher volumetric strain than the Tokimatsu-Seed approach as well as the approach described in present study with the exception from the depth greater than 7m. At such depths, a larger volumetric strain is predicted by the present approach. This is because the technique used in present study tends to overestimate slightly the volumetric strain at larger $(N_1)_{60}$ using equation (12.2).

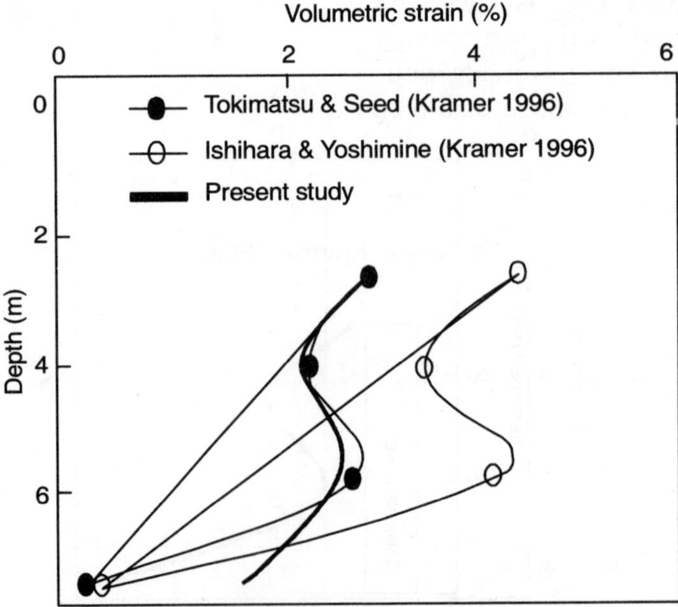

Fig. 12.11 Comparison of computed distributions of volumetric strain with depth (Courtesy: http://www.arpnjournals.com)

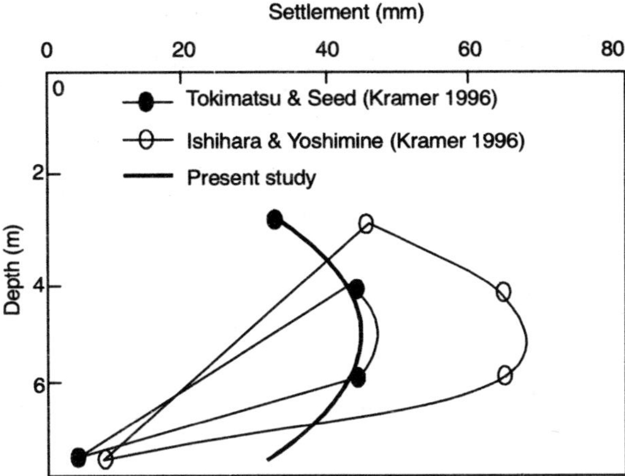

Fig. 12.12 Comparison of computed distributions of settlement with depth (Courtesy: http://www.arpnjournals.com)

Figure 12.12 shows the settlement variation with depth curves predicted by Tokimatsu-Seed, Ishihara-Yoshimine as well as the technique used in present study. The Ishihara-Yoshimine approach predicts higher settlement than those predicted by Tokimatsu-Seed as well as the technique used in present study. As expected, the technique used in present approach computes larger settlements at the depth greater than 7m due to the larger volumetric strains computed as shown in Figure 12.11.

Figure 12.13 compares the predicted performance using the Tokimatsu-Seed, Ishihara Yoshimine (reported in Kramer, 1996) and the technique used in present study with the measured settlement. The Tokimatsu-Seed approach under-predicts while the Ishihara-Yoshimine approach over-predicts the settlements. The settlement predicted by the technique used in present study compares favorably with the measurement. However, all the approaches are found to provide acceptable design prediction.

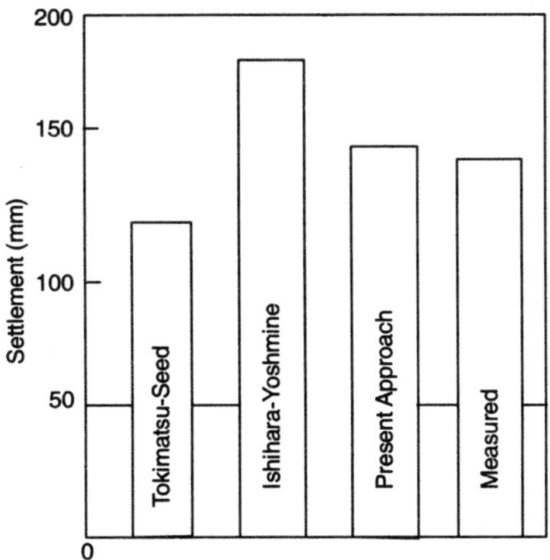

Fig. 12.13 Comparison of computed and measured settlements (Courtesy: http://www.arpnjournals.com)

12.4 CONCLUSIONS

A simple and practical approach has been presented in this chapter for evaluation of liquefaction-induced settlement in saturated sandy soils. The proposed approach described in this chapter does not require the use of charts, tables and diagrams. Consequently, it can be conveniently implemented numerically. Furthermore, it allows the expeditious analysis of multi-layered soil profile. The present approach may be able to characterize the liquefaction boundary line which separates liquefaction from non-liquefaction regions. It provides a simple way to explore the level of risk of liquefaction based on the cyclic stress and SPT values. The computed settlements by the technique described in present study is comparable with those by other published approaches. Reasonably good agreement is obtained between the predicted results by the technique of present approach and the measured results at two sites subjected to liquefaction-induced settlements. However, the technique used in present study is based on the limited data examined. Further validation of the technique presented in this chapter using additional field liquefaction data from earthquakes with various magnitudes is warranted.

Home Work Problems

1. Explain the following using simplified approach:

   (*i*) Volumetric strain after liquefaction.

   (*ii*) Magnitude scaling factor.

   (*iii*) Total settlement.

2. Compare volumetric strain and settlement variation with depth for Tokachi-Oki earthquake-Site P6 Hachinohe, Japan (1968) using Tokimatsu and Seed (1987) approach and the approach presented in this chapter.

13

CHAPTER

SITE INVESTIGATION FOR GEOTECHNICAL EARTHQUAKE ENGINEERING

13.1 INTRODUCTION

Type of facility to be constructed, nature of geologic hazard impacting site during earthquake, economic considerations, risk level and regulatory specifications are important factors affecting site investigation. Scope of site investigation for geotechnical earthquake engineering is usually divided into two parts: (1) screening investigation (2) quantitative evaluation of seismic hazards.

13.1.1 Screening Investigation

The purpose of screening investigation is to assess severity of seismic hazard at the site in question. Based on this, the sites which do not have seismic hazard are screened out. If site is free from seismic hazards, quantitative evaluation could be omitted. However, if site is likely to have seismic hazards, screening can define those hazards before proceeding with quantitative evaluation.

One important consideration for screening investigation is the effect that new construction will have on potential seismic hazard. When a screening investigation is carried out, both the existing condition and the final constructed condition must be evaluated for seismic hazards. Off-site seismic hazard (hazard on site due to earthquake at some distant location) is another important consideration. Screening investigation should be performed on regional as well as on site specific basis. Important steps are as follows:

Preliminary design information

Documents dealing with preliminary design as well as with proposed construction of project should be reviewed. Preliminary plan may even be developed showing proposed construction. For example structural engineer should have design information, such as building location, size, height, load and details on proposed construction materials and methods.

History of prior site development

It is important to obtain history of site if the site had prior development. Site might contain old deposits of fill, abandoned septic tanks and leach fields, buried storage tanks, seepage pits, cistens, mining shafts, tunnels etc. which might impact the proposed development. Old reports documenting seismic hazards at the site should also be reviewed.

Seismic history of area

Documents and maps providing data on seismic history of area should be studied. Seismic history information on the nature of past earthquake induced ground shaking should also be reviewed. This information could include period of vibration, ground acceleration, magnitude and intensity of past earthquake. This data can often be obtained from seismological maps and reports. These maps and reports are important because they can identify items such as pattern, type and movement of nearby active fault or fault systems as well as distance of fault to area under consideration. These records could also be reviewed to determine spatial and temporal distribution of historic earthquake epicenter.

Aerial photographs and geologic maps

During screening investigation, aerial photographs and geologic maps should be checked. They are useful in identifying existing and potential slope stability, fault ground rupture, liquefaction and other geologic hazards. Observed features include headwall scarps, debris chutes, fissures, grabens and sand boils. Comparison of older and newer aerial photographs helps in observing artificial or natural changes that have occurred at site with time. These reports and maps also indicate geometry of fault system, nature of fault system, sub-soil profile and seismic wave amplification due to local conditions. These all are important factors in the evaluation of seismic risk.

Special study maps

Special study maps have been developed to indicate local seismic hazards. They also indicate potentially liquefiable soil, landslides and abandoned mines.

Topographic maps

Old as well as recent topographic maps provide valuable site information. It shows the location of buildings, roads, freeways, train tracks, other civil engineering works, canyons, rivers, lagoons, sea cliffs, beaches etc. It might even show location of sewage disposal ponds and water tanks. Furthermore, by using different colors and shading it indicates older versus newer development. Main purpose of topographic map is to indicate ground surface elevation or elevation of the sea floor. This helps to determine major topographic features at site and to evaluate potential seismic hazards.

Regulatory specifications

Most recently adopted local building code should be reviewed. Design requirements for ordinary structures, critical facilities and lifelines may be delineated in the codes. They also provide seismic requirements for many buildings.

Other available documents

Other types of documents and maps are also useful during screening investigation. They include geologic and soil engineering maps and reports used for development of adjacent properties, water well logs, and agriculture soil survey reports.

After completing site research, field reconnaissance in screening investigation is carried out. Purpose is to observe site condition and document recent changes to site which may not be reflected in available documents. Field reconnaissance is also used to observe surface features and other details not evident from available documents.

13.1.2 Quantitative Evaluation

This is done to obtain adequate information on the nature and severity of seismic hazards. Consequently, adequate mitigation recommendations can be developed. Important steps are as follows:

Geologic mapping

Results of field reconnaissancs are supplemented with geologic mapping. This is then used to further identify such features as existing landslides as well as deposition of unstable soil on surface.

Subsurface exploration

Results of screening investigation and geologic mapping are used to plan subsurface exploration. It may consist of excavation of boring, test pit or trenches. Soil samples are often retrieved from excavation. Field tests are also performed in the excavations.

Laboratory testing

The purpose is to determine engineering properties of soil to be used in seismic hazard analyses.

Engineering and geologic analyses

Peak ground acceleration is important parameter for engineering and geologic analysis of seismic hazards.

Report preparation

Results of screening investigation and quantitative evaluation are presented in a report form. The report describes the seismic hazards and presents geologic as well as geotechnical recommendations.

13.2 SUBSURFACE EXPLORATION

Typically they consist of excavation of borings, test pits and trenches. Soil samples are retrieved from these excavations and then tested in laboratory to determine engineering properties. In addition field tests (standard penetration test or cone penetration test) can also be performed.

13.2.1 Borings, Test Pits & Trenches

Main objective of boring, test pits and trenches is to determine the nature and extent of seismic hazards. Their number and spacing for a particular project depends on judgement and experience. As their number increases, more knowledge is obtained about subsurface conditions and seismic hazards. This helps in creating more economical foundation design and less risk of project being impacted by geologic and seismic hazards.

If approximate idea of location of proposed structure is known, boring should be concentrated in that area. It could be drilled at four corners of a proposed building with additional one (and deepest) at the centre of proposed building. For unknown building location, boring is located in lines (such as across valley floor) to develop soil and geologic cross section.

If geologic and seismic hazard exists outside building footprint, they should be investigated with borings. If there is adjacent landslide or fault zone that could impact site, this should also be investigated with subsurface exploration. Factors affecting decision on number and spacing of boring is as follows:

Relative cost of investigation

Cost of additional boring should be weighed against value of additional subsurface information.

Project type

More detailed and extensive subsurface investigation is required for essential facility as compared to single family dwelling.

Topography

Hillside project requires more subsurface investigation than flatland project because of slope stability requirements.

Nature of soil deposit

If soil deposits are uniform compared to erratic deposits, fewer borings will be needed.

Geoplogic and seismic hazards

If there is greater potential for geologic and seismic hazard at site, greater will be need for subsurface exploration.

Access

If it becomes necessary to construct access road throughout site, it may be expensive, disruptive and may influence decision on number and spacing of borings.

Building requirements

For some projects, specifications are available for number and spacing of boring.

Initially preliminary subsurface plan involving limited number of borings is developed to have rough idea of soil, rock and groundwater conditions as well as potential geologic and seismic hazards at site. After analyzing preliminary subsurface data, additional boring as part of detailed seismic exploration is performed. This better defines soil profile, explores geologic and seismic hazards as well as obtains further data on critical subsurface conditions and seismic hazards.

Depth of excavation

Subsurface exploration should extend to a depth below which liquefiable soil is not expected to occur. At most sites, since soil is present boring or trench/test pit excavation is required. If bedrock is exposed over full site, simple surface inspection is sufficient. If local geology is well documented and ensures lack of liquefiable soil below exposed surface soil, then also simple surface inspection is sufficient.

Down hole logging

For landslides type of geologic hazards, large diameter bucket-auger boring that is down-hole logged is common form of subsurface exploration. It is valuable technique because it allows engineer to observe subsurface materials as they exist in place. Excavation of boring smears hole side and surface must be chipped away to observe intact soil or rock. Going down-hole is dangerous because of cave-in possibility of hole as well as presence of poisonous gases or lack of oxygen in the hole. Down-hole observation of soil and rock leads to discovery of important geologic and seismic hazards. Refer Fig. 13.1 and 13.2 for example.

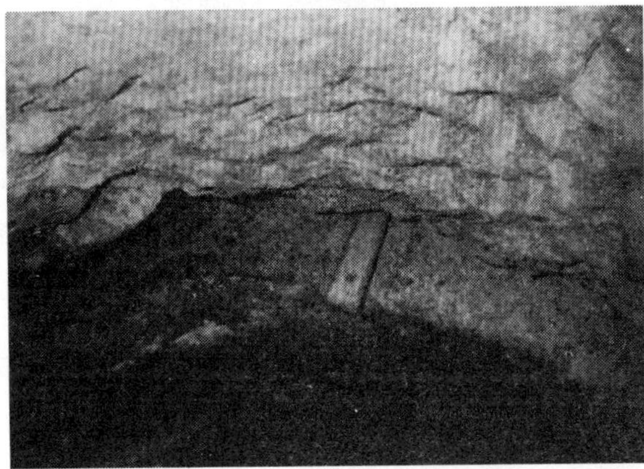

Fig. 13.1 Knife placed in an open fracture in bedrock caused by landslide movement (Courtesy: Day, 2002)

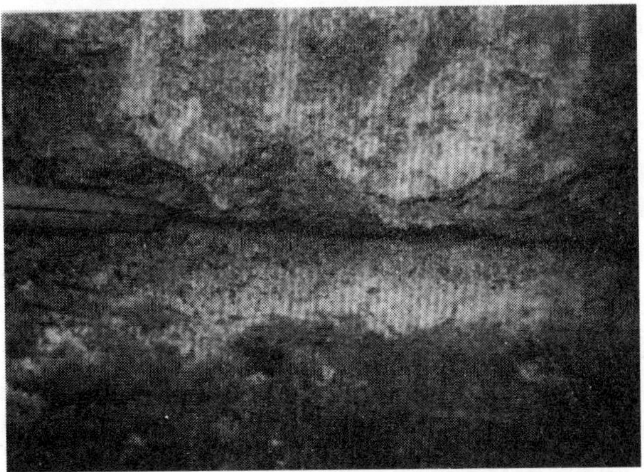

Fig. 13.2 Side view of the condition shown in Fig. 13.1 (Courtesy: Day, 2002)

Trench excavation

Backhoe trenches are economically used for this. Backhoe quickly excavates the trench, which is then used to observe and test in-situ soil. Backhoe trenches are used to evaluate near surface and geologic conditions (upto 15 feet deep), with borings to investigate deeper subsurface conditions. They are also useful for performing fault studies (Refer Fig. 13.3 and 13.4). Fault investigations are carried out to determine if there are active faults crossing the site. Width of shear zone of fault can also be determined from trench excavation studies. If activity of fault is uncertain, datable material present in trench excavation is used to determine date of most recent fault movement. Killed trees, sudden burial of marsh soils, disruption of archaeological sites are some of the datable materials.

Fig. 13.3 Backhole trench for a fault study (Courtesy: Day, 2002)

Fig. 13.4 Close-up view of trench excavation (Courtesy: Day, 2002)

13.2.2 SOIL SAMPLING

Ideally undisturbed soil sample should be obtained, field type stress conditions are applied and sample should be subjected to anticipated earthquake induced cyclic shear stresses. The resulting soil behavior could be used to evaluate seismic hazards. This requires undisturbed soil samples and sophisticated laboratory equipments. Since, the total process is expensive, other options as described here is used.

Cohesive soils

Unconfined compression test or consolidated undrained triaxial compression test is usually performed to determine undrained shear strength using laboratory analysis. Thin-walled Shelby tubes are used to obtain undisturbed cohesive soil samples.

Granular soils

Three techniques are available to obtain undisturbed soil sample in granular soils.

Tube sampling

Fixed piston sampler consists of piston fixed at bottom of bore hole by a rod extending to ground surface. Thin walled tube is then pushed into ground past the piston while piston rod is held fixed.

Sometimes groundwater table is lowered in the bore hole and water is allowed to drain from soil. This partially saturated soil is held together by capillarity enabling it to be sampled. Upon bringing it to the surface, the sample is frozen. Soil structure remains same in the process and frozen soil is transported to the laboratory. There is disruption of soil structure in the entire sampling process, but it is assumed as undisturbed. Greatest disturbance occurs during physical pushing of sampler in the soil.

Block sampling

For near surface soil, groundwater table is temporarily lowered making soil partially saturated. Then a test pit or trench is excavated in soil. A block sample is then cut from sides of test pit or trench and block sample is transported to laboratory for testing. Soil should have adequate capillarity to hold itself together. Soil could be disturbed due to stress relief when making excavation or when extracting soil sample.

Freezing technique

First soil is frozen and then sample is cut from ground. Freezing is accomplished by installing pipes in ground and then circulating ethanol along with dry crushed ice or liquid nitrogen through pipes. Slow freezing front should be established so that freezing water can expand and migrate out of soil pores minimizing sample disturbance.

The three techniques described above are not economical. Thus, laboratory technique is not practical and earthquake hazard analysis is based on field testing performed during subsurface exploration, most common being standard penetration and cone penetration testing techniques.

13.2.3 Standard Penetration Test

Although the test can be used for all soils, in general it should be used only for granular soils. The test consists of driving a thick walled sampler into granular soil deposit. The sampler has inside barrel diameter = 3.81cm and outside barrel diameter = 5.08cm as per ASTM D 1586–99 (2000). The sampler is driven into sand using 63.5kg hammer falling a distance of 0.76m. Sampler is driven a total of 45cm with number of blows recorded for each 15cm interval. The measured standard penetration test (SPT) N value is total number of blows required to drive sampler over depth interval of 15 to 45 cm.

The measured N value is influenced by soil type (amount of fines as well as gravel size particles) and groundwater. Saturated sand with lot of fines will give very high N value if they have tendency to dilate or very low N value if they have tendency to contract during sampler driving. Lot of gravel increases N value by becoming stuck in sampler tip. Level of water in borehole should be at or above in-situ ground water level. This will prevent groundwater from rushing into bottom of borehole decreasing measured N value.

Hammer efficiency, rate of blow application, borehole diameter and rod lengths are important testing factors affecting N value. Following equation is used (Skempton, 1986):

$$N_{60} = 1.67 E_m C_b C_r N \quad ...(13.1)$$

Where, N_{60} = standard penetration N value corrected for field testing.

E_m = hammer efficiency (= 0.6 for safety hammer and = 0.45 for doughnut hammer for US equipments).

C_b = borehole diameter correction (= 1 for 65–115mm diameter borehole, = 1.05 for 150mm diameter borehole and = 1.15 for 200mm diameter borehole).

C_r = rod length correction (= 0.75 for upto 4m drill rods, = 0.85 for 4 to 6m drill rod, = 0.95 for 6 to 10m drill rods and = 1 for drill rods in excess of 10m).

N = measured N value.

For many geotechnical earthquake engineering applications, the N_{60} value is corrected for vertical effective stress $\sigma'v_o$. Following equation is used:

$$(N_1)_{60} = C_N N_{60} = (100/\sigma'_{vo})^{0.5} N_{60} \quad ...(13.2)$$

Where, $(N_1)_{60}$ = N value corrected both for field testing and overburden pressure.

C_N = correction factor to account for overburden pressure (refer Fig. 13.5).

σ'_{vo} = vertical effective stress in kilo-pascals.

N_{60} = obtained from equation (13.1).

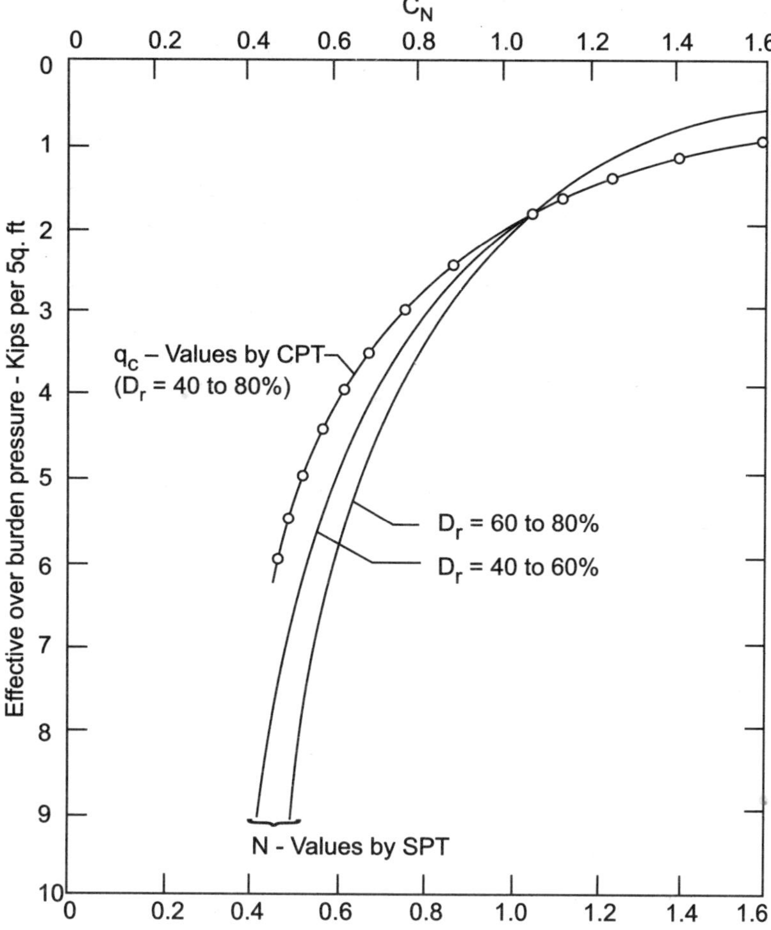

Fig. 13.5 C_N used to adjust N and q_c values. D_r indicates relative density (Courtesy: Day, 2002)

Some correlations are also available to estimate SPT value. SPT N value (blows per foot) and density condition of clean sand deposit is shown in Table 13.1. Very loose sand is subjected to significant settlement due to structure weight or due to earthquake shaking. Very dense sand is able to support high bearing loads and is resistant to settlement from earthquake shaking.

Table 13.1 Correlation between uncorrected SPT N value and density of clean sand (Courtesy: Day, 2002)

Uncorrected N value (blows per foot)	*Sand density*	*Relative density (%)*
0–4	Very loose	0–15
4–10	Loose	15–35
10–30	Medium	35–65
30–50	Dense	65–85
> 50	Very dense	85–100

Table 13.2 provides correlation between $(N_1)_{60}$ and relative density.

Table 13.2 Correlation between $(N_1)_{60}$ and Density of Sand (Courtesy: Day, 2002)

$(N_1)_{60}$ *(blows per foot)*	*Sand density*	*Relative density (%)*
0–2	Very loose	0–15
2–5	Loose	15–35
5–20	Medium	35–65
20–35	Dense	65–85
> 35	Very dense	85–100

Empirical correlation between N_{60}, vertical effective stress and friction angle of clean, quartz sand has been shown in Fig. 13.6.

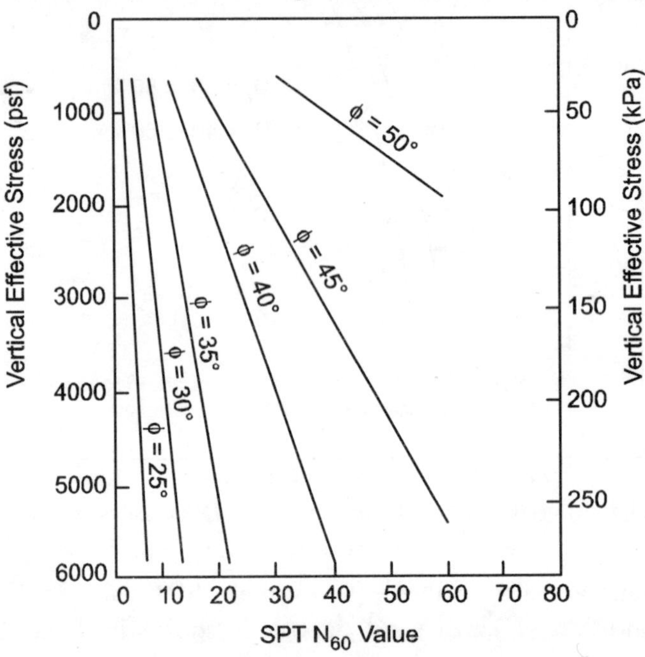

Fig. 13.6 Empirical correlation between N_{60}, vertical effective stress and friction angle for clean quartz sand (Courtesy: Day, 2002)

13.2.4 Cone Penetration Test

In this field testing, a steel cone is pushed into ground. Mechanical cone, mechanical-friction cone, electric cone and piezocone are different cone penetration devices used. First cone is pushed into soil to desired depth and then a force is applied to inner rods which moves cone downwards in extended position. Required force to move cone in extended position divided by horizontal projected area of cone gives cone resistance q_c. If this two-step process is repeated using electric cone (being pushed in soil at the rate of 10–20 mm/s), a continuous record of cone resistance versus depth is obtained. Fig. 13.7 presents empirical correlation between q_c, vertical effective stress and friction angle of clean, quartz sand.

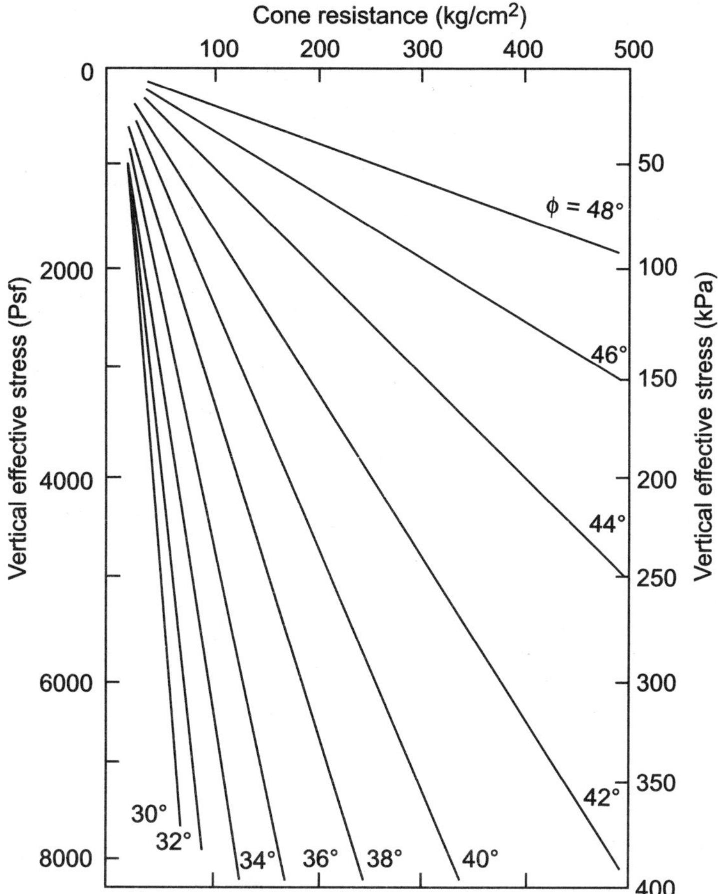

Fig. 13.7 Empirical correlation between cone resistance, vertical effective stress and friction angle for clean quartz sand deposit (Courtesy: Day, 2002)

For geotechnical earthquake engineering evaluations, q_c is corrected for vertical effective stress. Corrected q_c is called q_{c1}. Following equation is used:

$$q_{c1} = C_N q_C = \frac{1.8 q_c}{0.8 + \frac{\sigma'_{vo}}{100}} \qquad \ldots(13.3)$$

Where, q_{c1} = corrected CPT tip resistance.

C_N = correction factor to account for overburden pressure (refer Fig. 13.5).

σ'_{vo} = vertical effective stress in kilopascals.

Using electric cone, continuous subsurface record of q_c is obtained. However, in standard penetration test, data at intervals in soil deposit is obtained. But it is difficult to obtain slow and steady penetration of cone.

13.3 LABORATORY TESTING

Special laboratory tests used to model engineering behavior of soil subjected to earthquake loading is typically not performed in practice. The field case studies and in-situ tests provide most useful and practical tool.

Shear strength is important parameter needed for earthquake analyses of foundations, slopes and retaining walls. Cyclic triaxial test is used to study dynamic properties of soil.

13.3.1 Shear Strength

Two basic types of analyses utilize shear strength of soil: (1) total stress analysis (2) effective stress analysis. They can't be combined under any circumstances. Suppose factor of safety is to be determined for slope consisting of alternate sand and clay layers. As per total stress analysis, total shear strength parameters will be used, total unit weight of soil will be used and groundwater table will be ignored. As per effective stress analysis, effective shear strength parameters will be used and earthquake induced pore pressure will be determined.

Total stress analysis

It is performed for silts and clays. Undrained shear strength of soil is used. It is used for design of foundations, slopes and retaining walls subjected to earthquake shaking. During earthquake, there is rapid loading or unloading, there is quick change in shear stress and soft cohesive soil has no time to consolidate as well as negative pore pressure in heavily overconsolidated cohesive soil has no time to dissipate. Total unit weight of soil is used and groundwater location is not considered in analysis.

Undrained shear strength of soil is determined using unconfined compression or vane shear tests. Sometimes total stress parameters from unconsolidated undrained triaxial compression test or consolidated undrained triaxial compression test are also used.

Although total stress analysis is easy, accuracy of undrained shear strength is always in doubt, primarily because it depends on shear induced pore pressures (not measured), which in turn depends on many details.

Effective stress analysis

Earthquake analysis of granular soils make use of effective stress analysis. Effective cohesion is zero for such soils and effective friction angle is obtained from drained direct shear tests or from empirical correlations such as standard penetration or cone penetration tests.

For earthquake induced loading, this analysis requires estimation of earthquake induced pore pressures or effective stress generated during earthquake must be determined. It is more fundamental because shear strength is directly related to effective stress. But, accuracy of pore water (to be used in analysis) is doubtful as many parameters during earthquake loading affect it.

Cohesionless soil

This kind of soil can be held together only by confining pressure and will fall apart when confining pressure is released. It is easier to perform effective stress analysis for such soils under earthquake conditions due to variety of reasons.

Cohesionless soil above groundwater table has negative pore pressure due to capillary tension in pore fluid. Capillary tension provides additional shear strength to soil as particles are held together. Similar to geotechnical analysis, in earthquake analysis also capillary tension is ignored. Shear strength of soil above groundwater table is assumed to be equal to effective friction angle. The effective friction angle is obtained from empirical correlations (Figs. 13.6 & 13.7) or from drained direct shear tests performed on saturated samples.

Dense cohesionless soil below groundwater table tends to dilate during earthquake shaking. Excess pore pressure becomes negative, and soil shear strength momentarily increases. Thus for dense cohesionless soil below groundwater table, shear strength is equal to effective friction angle, obtained from empirical correlations (Figs. 13.6 & 13.7) or from drained direct shear tests performed on saturated samples. In effective stress analysis, negative excess pore pressure is ignored and pore pressure is taken as hydrostatic.

Loose cohesionless soil tends to contract during earthquake shaking. Pore pressure induces and shear strength decreases. If liquefaction of this soil occurs, shear strength becomes zero. Analysis should incorporate reduction in shear strength due to increase in earthquake induced pore pressure as soil contracts, if factor of safety against liquefaction is greater than 1. It is very difficult to estimate pore pressure induced due to earthquake induced contraction of soil. Sometimes pore water pressure ratio r_u is obtained as a function of factor of safety against liquefaction FS_L, from Fig. 13.8. $r_u = u/(\gamma_t h)$, where, u = pore water pressure, γ_t = total unit weight of soil & h = depth below ground surface. If $r_u = 1$, $u = \gamma_t h$, and effective stress = 0.

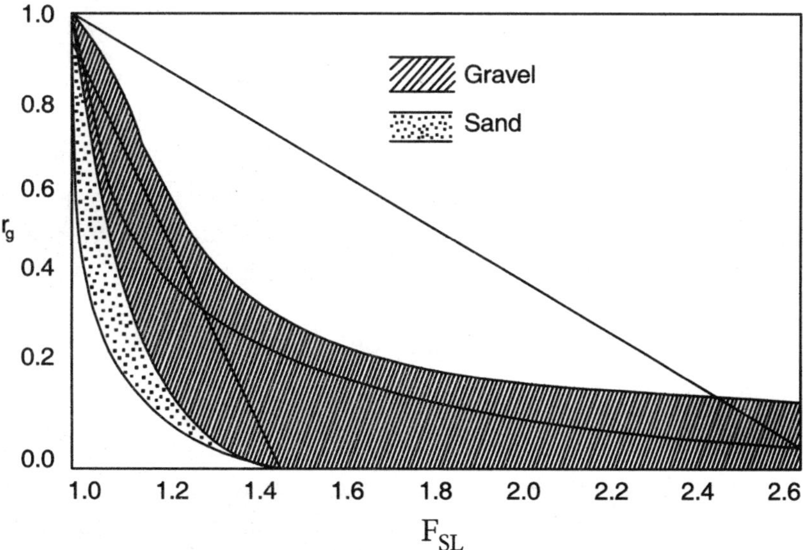

Fig. 13.8 FS_L versus pore water pressure ratio for gravel and sand (Courtesy: Day, 2002)

Flow failure cases in cohesionless soils is analyzed using effective stress analysis with $r_u = 1$, or by using shear strength of liquefied soil equal to zero ($\phi' = 0$ & $c' = 0$).

Cohesive soil

These soils are plastic and have ability to be rolled and molded. For earthquake analysis total stress analysis should be performed on these soils.

Cohesive soil above groundwater table has negative pore pressure due to capillary tension of pore fluid. Soil may even be dry and desiccated. Due to capillary tension particles are held together, providing additional shear strength. Undrained shear strength can be determined from unconfined compression or vane shear tests. Otherwise, total stress parameters from triaxial tests can be determined. Due to negative pore pressure, future increase in water content decreases undrained shear strength of partially saturated such soil and this increase should be included in analysis. If triaxial test on partially saturated cohesive soil shows significant difference between peak and ultimate value of stress strain curve, ultimate value should be used for earthquake analysis.

Earthquake tends to shear cohesive soil back and forth. For cohesive soil below groundwater table having low sensitivity (value ≤ 4), reduction in shear strength during earthquake is small.

If cohesive soil below groundwater table has high sensitivity (value > 8), there will be significant shear strength loss due to earthquake shaking. There is significant difference between peak and ultimate value of stress-strain curve during triaxial testing and this difference should be included in the analysis for such soils. If during earthquake, sum of static and seismic induced shear stresses exceed undrained shear strength, significant reduction in shear strength occurs.

Cohesive soil having medium sensitivity has intermediate behavior.

Stability analysis of ancient landslides, slopes in overconsolidated fissured clays and slopes in fissured shales are examples of cohesive slope subjected to significant shear deformation. For stability analysis of such slopes during earthquake shaking, drained residual shear strength should be used. This is determined using torsional ring shear or direct shear device. While performing effective stress analysis, pore pressure is assumed unchanged. Slope is also subjected to additional destabilizing forces due to earthquake shaking. These forces are included in effective stress slope stability analysis (pseudostatic method).

Subsoil profile consisting of cohesive and cohesionless soil

If both cohesionless and cohesive soil has to be considered, either total or effective stress analysis could be performed. For cohesionless soil, effective strength parameters are known. Hence, subsoil having layers of sand and clay is analyzed using effective stress analysis, with estimation of earthquake induced pore pressure. If sand layer will liquefy during earthquake, total stress analysis should be performed using undrained shear strength for clay and assuming undrained shear strength of sand layer = 0. Bearing capacity or slope stability analysis using total stress parameters can be performed such that failure surface passes through liquefied sand layer.

13.3.2 Cyclic Triaxial Test

It has been used extensively to study soil subjected to simulated earthquake loading. A cylindrical soil sample is placed in triaxial apparatus and sealed in watertight rubber membrane. Sample is saturated using back pressure. Isotropic effective confining pressure is applied to sample and it is allowed to equilibrate. Tubing is provided which allows water flow during saturation and equilibration as well as pore pressure measurement (refer Fig. 13.9). At the end of equilibration, valve to drainage measurement is shut and sample is subjected to undrained loading. Constant amplitude sinusoidally varying axial load (cyclic axial load) is applied on top of sample to simulate earthquake loading. Cyclic axial load simulates shear stress change induced due to earthquake. Cyclic axial load, sample axial deformation and sample pore water pressure is measured. Cyclic axial load in loose sand increases pore pressure, decreases effective stress and increases axial deformation. It is very complicated test, requiring special equipment and many factors affect the result.

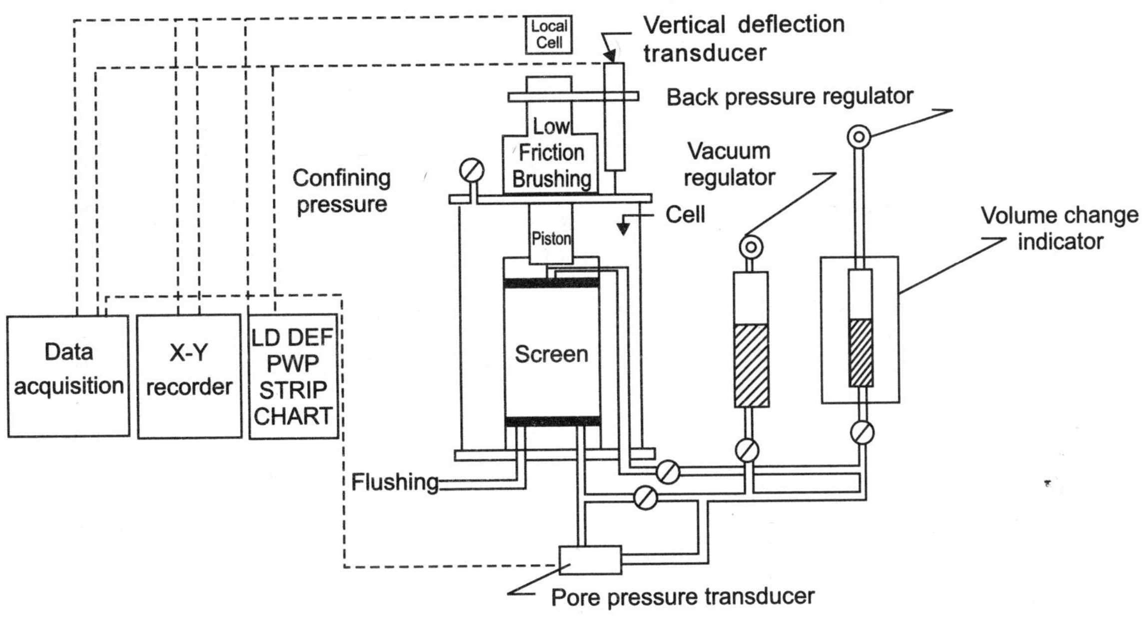

Fig. 13.9 Schematic diagram of cyclic triaxial test equipment (Courtesy: Day, 2002)

13.4 PEAK GROUND ACCELERATION

Ground surface acceleration during earthquake is directly related to dynamic forces induced on soil mass and hence it is measured. Cyclic ground motion is represented by maximum horizontal acceleration at ground surface a_{max}. It is also called peak horizontal ground acceleration. Usually horizontal acceleration is more than vertical acceleration, hence peak horizontal ground acceleration is also called peak ground acceleration (PGA). a_{max} represents acceleration which will be induced sometimes in future by earthquake. Earthquakes can't be predicted, so PGA is based on prior earthquakes or fault studies.

Attenuation relation (mathematical relation to estimate PGA at specified distance from earthquake) is used to estimate PGA. These relations relate peak ground acceleration to earthquake magnitude as well as distance between site and seismic source. Empirical attenuation equations are also available to model ground motions during earthquake.

13.4.1 Methods Used to Determine PGA

Engineering geologist is best individual to determine PGA based on fault, seismicity and attenuation relations. Commonly used methods are as follows:

Historical earthquake

Past earthquake history of site is considered. For recent earthquakes seismograph data can be used to determine PGA. For older earthquakes, location and magnitude is based on historical damage account. Computer program EQSEARCH has been developed to incorporate past earthquake data. Location of site is given as input and a_{max} is determined. However, historical time frame of recorded earthquake is too small and a_{max} should never be based on history of seismic activity in the area alone. It should be compared with other methods also.

Regulatory requirements

There are regulatory requirements specifying design value of PGA. A given area is divided into different seismic zones. Peak ground acceleration coefficient (a_{max}/g) has standard value in each zone. However, depending on distance to active fault and subsoil profile, this standard value may underestimate or overestimate PGA.

Maximum credible earthquake

It is largest earthquake that can reasonably be expected to occur based on known geologic and seismic data. It is maximum earthquake an active fault can produce, considering geologic evidence of past movement and recorded seismic history. It is determined for a given level of ground shaking. The analysis used to determine maximum credible earthquake is called deterministic method.

Maximum probable earthquake

It is based on study of nearby active faults. Using attenuation relation, maximum probable earthquake magnitude and maximum probable peak ground acceleration can be determined. Maximum probable earthquake is largest predicted earthquake, a fault is capable of generating within specified time period such as 50 or 100 years. It is most likely to occur during design life of project and has been commonly used in assessing seismic risk. Sometimes it is also defined as earthquake that will produce peak ground acceleration a_{max} with 50% probability of exceedance in 50 years.

Earthquake maps

Design basis ground motion is determined by site specific hazard analysis or from hazard map. Usually these maps show peak ground acceleration with certain probability of exceedence in 50 years and provide user with choice of appropriate level of hazard or risk. The approach is probabilistic method with choice of peak ground acceleration based on concept of acceptable risk.

There could be considerable variation in value of a_{max} obtained from different methods and geotechnical engineer should work with engineering geologist in selecting most appropriate value of a_{max} for a particular project at a particular site.

13.4.2 Local Soil and Geologic Conditions

Final step in determination of a_{max} for particular site is to adjust the obtained value for different factors. They are as follows:

— directivity of ground motion, causing stronger shaking in certain directions.

— Thick deposit of soft soil, which increases peak ground acceleration.

— Basin effect, (conversion to surface wave and reverberation experienced by site in alluvial basin).

Example 13.1 A standard penetration test (SPT) was performed on near surface deposit of clean sand, where number of blows to drive sampler 18in was 5 for first 6in, 8 for next 6in and 9 for third 6in. Calculate measured SPT N value (blows per foot) and indicate density condition of sand as per Table 13.1.

Solution:

N value = 8 + 9 = 17, as per Table 13.1, sand has medium density.

Example 13.2 A clean sand deposit has level ground surface, total unit weight above groundwater table 18.9 kN/m^3, and submerged unit weight 9.84 kN/m^3. Groundwater table is 1.5m below ground surface. SPT was conducted in 10cm diameter bore hole. At depth of 3m below ground surface, SPT performed using doughnut hammer had blow count of 3 for first 6in, 4 for second 6in and 5 for third 6in of driving penetration. Assuming hydrostatic pore pressure, determine vertical effective stress and corrected N value (corrected for field testing) at 3m depth.

Solution:

Effective stress $= \gamma_t\ (1.5) + \gamma_b\ (1.5) = 18.9(1.5) + 9.84(1.5) = 43$ kPa.

$N_{60} = 1.67 E_m C_b C_r N$

Where, $E_m = 0.45$ (doughnut hammer), $C_b = 1.0$ (100mm diameter hole), $C_r = 0.75$ (3m length of drill rods), $N = 4+5 = 9$. Substituting these values,

$N_{60} = 1.67(0.45)(1.0)(0.75)(9) = 5.1$

Home Work Problems

1. Using data from Example 13.2, determine N value corrected for both field testing, and overburden pressure, and indicate in-situ condition of sand as per Table 13.2.

(**Ans**. Corrected N value = 7.8, sand has medium density).

2. Use data from Example 13.2, assume cone penetration test was performed at depth of 3m and cone resistance = 40 kg/cm^2. Determine CPT tip resistance corrected for overburden pressure.

   (**Ans.** CPT tip resistance corrected for overburden pressure = 59 kg/cm^2).

3. Explain in detail scope of site investigation for geotechnical earthquake engineering.
4. Write short notes on:

   (*i*) Standard penetration test

   (*ii*) Cone penetration test
5. What are two basic types of analyses that utilize shear strength of soil. Explain them.
6. Draw labeled diagram of cyclic triaxial test equipment.
7. What are the methods to determine peak ground acceleration. Explain them.

14

CHAPTER

GEOTECHNICAL EARTHQUAKE ENGINEERING FOR RIVER BRIDGE: A CASE STUDY

14.1 INTRODUCTION

The proposed Great River Bridge is going to span the Mississippi River from Desha County, Arkansas to Bolivar County, Mississippi (refer Fig. 14.1). HNTB Corporation is a leading team of consultants in the design of the bridge. URS Corporation, a member of the HNTB team, performed the geotechnical earthquake engineering for the bridge.

The approach structure on the Arkansas side is approximately 14,680 feet long. This length is from the beginning of the bridge to the west bank main span anchor pier. Important crossings in Arkansas include the U.S. Corps of Engineers' (USCOE) Arkansas levee and a wing dam. The proposed main structure is a steel cable-stay structure. This structure has a navigation span of approximately 1400 feet (refer Fig. 14.2). The approach structure on the Mississippi side is approximately 2,610 feet in length. This length is from the east bank main span anchor pier to the end of the bridge. The USCOE Mississippi levee is the only important crossing in Mississippi. The total length of the project is approximately 23,500 feet. Out of this, approximately 22,548 feet length is of bridge.

The bridge site is located approximately 120 miles south of the New Madrid Seismic Zone. Equivalent-linear ground response analysis was used to propagate ground motions through the near-surface soils. Resulting near-surface motions was used to evaluate liquefaction, earthquake-induced settlements, and seismic slope stability. Analysis indicated that liquefaction would occur across the site and would result in liquefaction-induced settlement, lateral spreading at the Arkansas riverbank, and seismic deformations of proposed and existing embankments. To reduce liquefaction-related design issues, several mitigation options were investigated. The: (1)Field and laboratory investigation; (2)Geologic and seismologic setting; (3)Evaluation of ground shaking, site response, liquefaction, and seismic slope stability; (4)Earthquake-related foundation recommendations; and (5)Conceptual liquefaction mitigation options have been described in detail in this chapter.

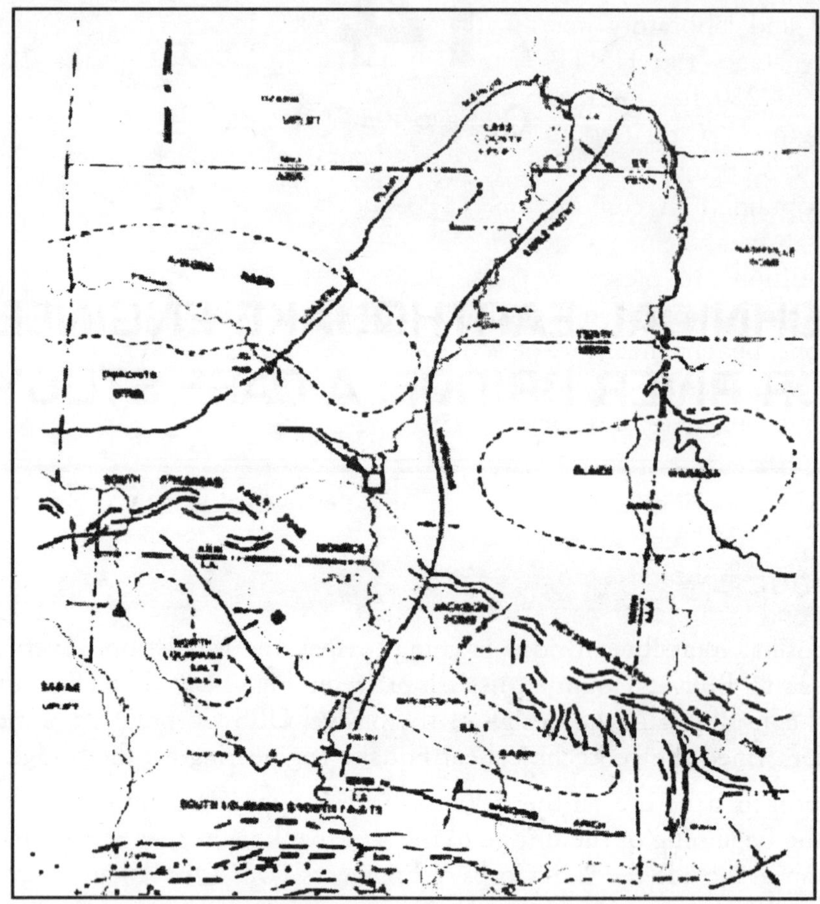

Fig.14.1 Project location and geologic structures of Mississippi embayment (Courtesy: https://netfiles.uiuc.edu)

Fig. 14.2 Proposed great river bridge (Courtesy: https://netfiles.uiuc.edu)

14.2 FIELD AND LABORATORY INVESTIGATION

The field and laboratory investigation was divided into preliminary and final stage. For preliminary stage, the HNTB team drilled 15 exploratory borings in Arkansas to depths of 150 to 220 feet and 5 borings in Mississippi to depths of 120 to 220 feet. Rotary wash methods were used to advance each of the borings. Split spoon and Shelby tube soil samples were obtained at various depths in each of the borings. The split spoon samples were obtained in conjunction with standard penetration tests using automatic hammers. Energy measurements were made for each of the automatic hammers used at the site. In addition, 36 piezocone penetration test soundings, 26 in Arkansas and 10 in Mississippi were performed in accordance with specifications. In six of these soundings, seismic piezocone penetrometer was used to obtain shear wave velocity measurements at one-meter intervals. Lastly, downhole geophysical testing to depths of 150 feet in four boreholes were performed to measure compressive and shear wave velocities.

The final field investigation consisted of 67 borings in Arkansas, 19 borings in Mississippi, and 12 borings in the river performed from a barge. In addition, 18 more piezocone penetration test soundings were performed to better delineate subsurface conditions in the approach embankment area well as near the Arkansas levee. The HNTB team conducted laboratory testing on selected soil samples obtained during the preliminary and final geotechnical investigations. These tests included index, consolidation, and strength tests.

14.3 GEOLOGIC SETTING

The project site lies within the Mississippi Embayment, part of the Atlantic and Gulf Coastal Plains. The Embayment fills a southward-plunging syncline that extends from the southern tip of Illinois to the Gulf of Mexico. It is located between the Ozark Uplift and the Nashville Dome (Ervin and McGinnis 1975). The axis of the syncline roughly follows the present course of the Mississippi River (Cushing et al. 1964). The Embayment dates to the Paleozoic period. During this period, crustal movement and compaction of deposits caused subsidence and created a basin for the deposition of alluvial sediments (Cushing et al. 1964; Romero and Rix 2001).

The surface geology at the site consists of Quaternary (Pleistocene and Holocene) alluvium overlying a thick column of sedimentary deposits of varying induration ranging in age from Jurassic to Quaternary. Paleozoic basement rock was encountered at about El. –4700 (feet, mean sea level datum). The Jackson Group is hard, tertiary-age clay encountered approximately 100 to 140 feet below grade at the project site. The Jackson Group was deposited during the last widespread inundation by the sea. Since the tertiary period, most of the Mississippi Embayment has remained above sea level (Cushing et al. 1964; Saucier 1994). During the Quaternary, the alluvial fill of the Mississippi River was deposited and extensive terrace systems were formed as a result of changing sea levels (Saucier 1994).

14.4 SEISMIC SETTING

Seismic hazard near the project site is both from potential local earthquakes and from larger events occurring farther away. Of particular significance is the New Madrid Seismic Zone to the north. Damaging earthquakes were also originated in Arkansas and Mississippi outside the New Madrid Seismic Zone. A number of earthquakes with magnitudes between 4 and 5 were documented in southern Arkansas and central Mississippi.

The New Madrid sequence of 1811–12 contained at least three primary earthquakes. The exact locations of the earthquakes are uncertain. Reason being the lack of instrumental data and the sparse local population during the period. Nuttli (1973), on the basis of variations in the spatial extent of the most serious damage and felt effects, suggested that the first event occurred in northeastern Arkansas, the second between New Madrid, Missouri and the location of the first event, and the third near New Madrid (Fig. 14.3). Numerous aftershocks occurred with diminishing frequency for more than 5 years after the three main earthquakes (Nuttli 1982). Recent estimates place the moment magnitude of these earthquakes on the order of 7.6 or 7.7 ± 0.5 (Wheeler and Perkins 2000; Atkinson et al. 2000). Current monitoring shows this zone is still active as shown in Fig. 14.3.

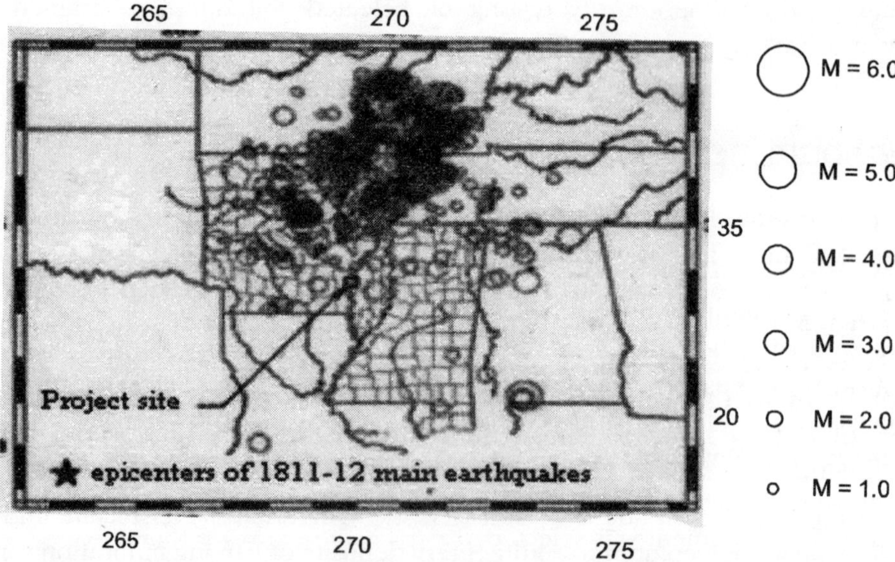

Fig. 14.3. Earthquakes from 1974 through May 2002 within 500 km of Site (Courtesy: http://www.ceri.memphis.edu)

14.5 SITE AND SUBSURFACE CONDITIONS

On both sides of the Mississippi River, the site generally is flat. It consists of agricultural fields, backswamps, and wooded areas.

It has already been explained that the upper strata at the site consisted of Holocene and Pleistocene alluvial deposits to depths of about 100 to 140 feet overlying Tertiary-age clays. The alluvial deposits can be separated into four distinct geologic categories: overbank silts and clays; abandoned channel (i.e., cutoff meander) deposits; point bar clean sands; and braided stream sand deposits. The abandoned channel can be subdivided into alluvial and lacustrine clays and silty/sandy channel deposits. The braided stream deposits can be subdivided into medium dense alluvial sands and dense alluvial sands and gravelly sands. Table 14.1 provides in situ test results for the various strata.

Underlying the alluvial deposits is a Tertiary-age, hard, high plasticity clay with occasional layers of low plasticity silty clay, sandy clay and silt. Most likely, the hard clays present at the site belong to the Yazoo Formation of the Jackson Group. Table 14.1 includes in situ test results for this stratum.

Table 14.1 In situ test results for subsurface materials (Courtesy: https://netfiles.uiuc.edu)

Stratum	*SPT blow count (bpf)*	*CPT tip resistance (tsf)*	*Shear wave velocity (ft/sec)*
Over-bank clay	1–12	10–70	600
Channel silt/sand	2–28	20–200	600
Channel clay	0–12	5–40	450
Point bar sand	3–17	40–300	600
Med. dense sand	5–47	100–300	850
Dense sand	10–100$^+$	200–300$^+$	1250
Tertiary clay	19–100$^+$	n/a	1400

Groundwater readings were taken in each of the completed boreholes. Furthermore, piezometers were installed in seven boreholes at various depths to monitor groundwater conditions away from the river. Measurements indicated that the groundwater level ranged from 10 to 20 feet below grade and dropped toward the river. Groundwater levels near the river closely mirrored river level changes.

For the geotechnical earthquake engineering analyses, a design groundwater level at El. 125 along the entire alignment was selected. This elevation is lower than the groundwater level corresponding to the 5-year flood stage, El. 138. The low water reference plane is El. 99.5. A design groundwater level of El. 125 was selected to provide a reasonable groundwater level considering the severity of the earthquake loading as well as the low probability that the design earthquake would occur during high river stage of a 5-year flood.

14.6 GROUND SHAKING

Based on project design criteria established by the state of Arkansas, bedrock shaking at the project site was evaluated based on seismic design parameters recommended by the 1996 USGS (united states geologic survey) seismic hazard maps (Frankel et al. 1996). These seismic parameters were used to select and modify site-specific bedrock acceleration time histories. These bedrock time histories were then used to perform ground response analyses based on soil conditions encountered at the site.

14.6.1 Bedrock Spectrum and Time Histories

The bedrock response spectrum from the USGS national seismic hazard maps for 2% probability of exceedence in 50-years is shown in Fig. 14.5. This response spectrum is for Profile – 1, general conditions. Details about Profile – 1 is given in the first paragraph of section 14.6.2. Somerville et al. (2001) suggested that horizontal and vertical bedrock spectra for Eastern North America (ENA) are nearly identical. On this basis, the bedrock response spectrum shown in Fig. 14.5 corresponds to both horizontal and vertical bedrock ground motions. For later analyses, the probabilistic response spectrum was deaggregated and it was concluded that large distant earthquakes occurring in the NMSZ (New Madrid Seismic Zone) dominated the ground motions at the site. The deaggregated moment magnitude (M_W)-distance pair was found to correspond to approximately 8 and 200 km, respectively.

Table 14.2 lists the bedrock acceleration time histories selected for the Great River Bridge. The first time history set was from 1988 Saguenay, Quebec earthquake (M_W = 5.8; Somerville et al. 1990). The Saguenay earthquake is the largest earthquake in eastern North America for which strong motion recordings have been obtained. The other two time history sets are from broadband strong motion simulations performed by Somerville et al. (2001). These time histories are representative of M_W = 7.5 earthquakes occurring on various segments of the New Madrid fault system at a distance of 75 miles. The selected time histories provide comparable peak accelerations, velocities, and displacements to those anticipated for the deaggregated magnitude-distance pair applicable to the Great River Bridge.

Table 14.2 Time histories selected for analysis (Courtesy: https://netfiles.uiuc.edu)

Source	M_w	*Distance (miles)*	*Components*
1988 Saguenay earthquake	5.8	27	124, 243, vertical
New Madrid A (deep hypocenter)	7.5	75	East, north, vertical
New Madrid B (shallow hypocenter)	7.5	75	East, north, vertical

The acceleration time histories in Table 14.2 do not correspond precisely to the bedrock response spectrum. Therefore prior to conducting ground response analyses, these time histories were spectrally-matched to the bedrock (target) spectrum using the method proposed by Lilhanand and Tseng (1988) as modified by Abrahamson (1993). The 5% damped, spectrally-matched bedrock motions were found to closely agree with the USGS probabilistic bedrock response spectrum.

14.6.2 Ground Response Analysis

To simplify the ground response analysis and provide a means to differentiate various reaches along the 4.3–mile long project alignment, the average shear wave velocity to the top of the Jackson Group at discrete stations along the alignment as shown in Fig. 14.4 was calculated. Ground response analysis for 3 representative soil columns that had average V_s (shear wave velocity) values that fell within the ranges shown in Fig. 14.4 was conducted. These representative soil columns were labeled as: Profile 1—General Conditions; Profile 2 — Abandoned Channel; and Profile 3 — Riverbanks & Point Bar.

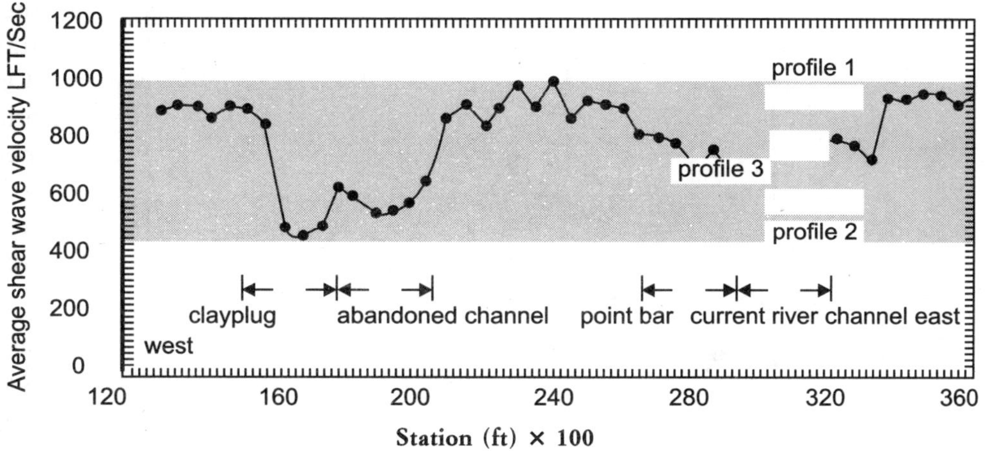

Fig. 14.4 Average shear wave velocity to top of Jackson group (Courtesy: https://netfiles.uiuc.edu)

The bedrock motions developed for the site corresponded to a moderately hard rock site with V_s=2500 ft/sec. The shear wave velocity profile was extended to the depths of 200–500 feet using limited existing geophysical measurements available in the New Madrid Seismic Zone (as summarized in Romero and Rix 2001).

Both horizontal and vertical ground response analysis was performed. However, only the horizontal analysis has been described here. Cooling et al. (2003) described the results of the horizontal and vertical ground response analyses in detail. One-dimensional, equivalent-linear ground response analyses using the computer software SHAKE (Schnabel et al. 1972), as most recently updated in 1996 was carried out. This analysis assumes that horizontal motions can be approximated using vertically propagating shear waves (S-waves). Fig. 14.5 presents the horizontal ground response results for Profile 1 - General conditions. Analysis at each soil profile included runs for each horizontal component of each earthquake (i.e., total of 6 time histories) for the "shallow" soil column, the "deep" soil column, and parametric runs using the "shallow" soil column with ± 20% of the average shear wave velocity.

Using the ground response analyses for the individual soil profiles, smoothed, design horizontal response spectra for each profile was developed as summarized in Fig 14.6. As anticipated, the lower shear wave velocity profile yielded a softer response (i.e., lower accelerations at short periods and higher accelerations at long periods), while the higher shear wave velocity profile yielded a stiffer response (i.e., higher accelerations at short periods and lower accelerations at long periods). The ground motions correspond to the top of the pile caps for each foundation group (i.e., approximately 10 feet below grade) and do not include the effects of liquefaction. Using these smoothed design response spectra, 27 spectrally matched acceleration time histories at the top of the pile caps were calculated. The 27 time histories consist of 3 orthogonal components (x, y, and z) × 3 soil profiles × 3 earthquakes.

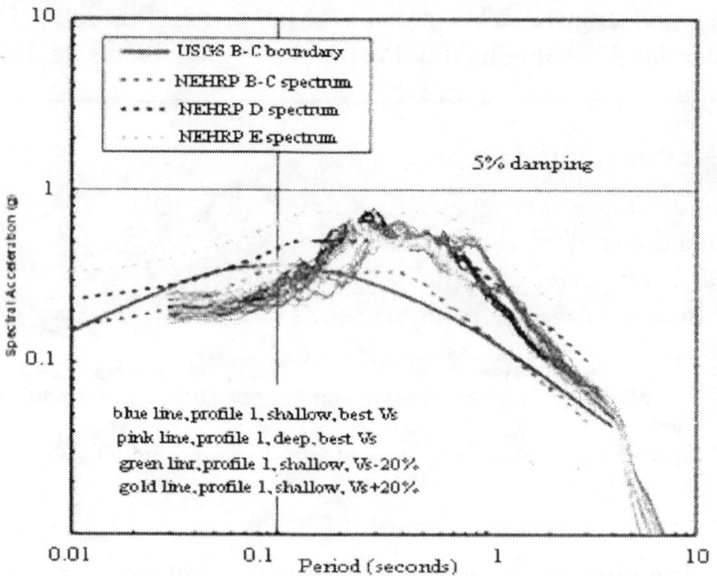

Fig. 14.5 Response spectra for profile-1, general conditions (Courtesy: https://netfiles.uiuc.edu)

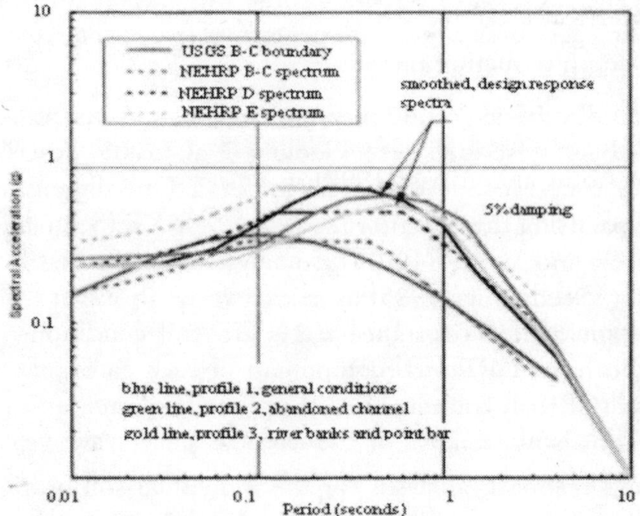

Fig. 14.6 Summary of horizontal design response spectra (Courtesy: https//netfiles.uiuc.edu)

14.7 LIQUEFACTION

Liquefaction is the process of strength loss in saturated, cohesionless soils as a result of an increase in porewater pressure. Often it is due to earthquake shaking. Liquefaction can have numerous detrimental effects on natural and man-made structures. This includes settlement, bearing capacity failure, downdrag on deep foundation elements, lateral spreading, and large-scale slope instability. This section describes the analysis of liquefaction under level or mildly sloping (slopes < 6%) ground conditions. The effects of liquefaction on slope stability has been discussed subsequently.

14.7.1 Level Ground Liquefaction Analysis

Level ground liquefaction analysis compares the load imposed by earthquake shaking (also called seismic demand) to the resistance of the soil to pore water pressure increase (also called seismic capacity). Seed and Idriss (1971) developed the "cyclic stress" method to estimate seismic demand in terms of the approximate seismic stresses imposed as a result of earthquake shaking. The cyclic stress ratio (CSR) is defined as:

$$CSR = \frac{\tau_{avg,cyclic}}{\sigma'_v} = 0.65\frac{\tau_{max,cyclic}}{\sigma'_v} = 0.65\left(\frac{a_{max}}{g}\right)\left(\frac{\sigma_v}{\sigma'_v}\right)r_d \qquad ...(14.1)$$

Where,

τ_{cyclic} = approximate seismic shear stress.

a_{max} = free-field peak ground acceleration.

σ_v = vertical total stress.

$\sigma'v$ = vertical effective stress.

r_d = depth reduction factor.

Horizontal peak ground acceleration of 0.26g for the abandoned channel and 0.23g for the remainder of the site was used. It was based on the ground response analysis results. Values of r_d as proposed by Youd and Idriss (1997) was used.

The seismic capacity, or liquefaction resistance, was defined using the CPT procedures proposed by Robertson and Wride (1997) and Stark and Olson (1995) as well as the SPT procedure proposed by Seed et al. (1985) as modified by Youd and Idriss (1997). Shear wave velocity to evaluate liquefaction resistance was not used due to minor difficulties in the geophysical testing. Both the CPT and SPT procedures define liquefaction resistance using a cyclic resistance ratio (CRR) for a magnitude 7.5 earthquake ($CRR_{7.5}$). The factor of safety against liquefaction (FS) is written as:

$$FS = \left(\frac{CRR_{7.5}}{CSR}\right)MSF \qquad ...(14.2)$$

MSF is a magnitude scaling factor in equation (14.2). It adjusts the value of $CRR_{7.5}$ to earthquake magnitudes other than 7.5. Since the de-aggregated earthquake magnitude is

M_W 8.0, the value of MSF was taken as 0.88. Using the CPT and SPT results, (Fig. 14.7) shows zones of probable liquefaction (FS≤1) along the project alignment. Liquefaction was found to be most prevalent within the abandoned channel and in the current point bar.

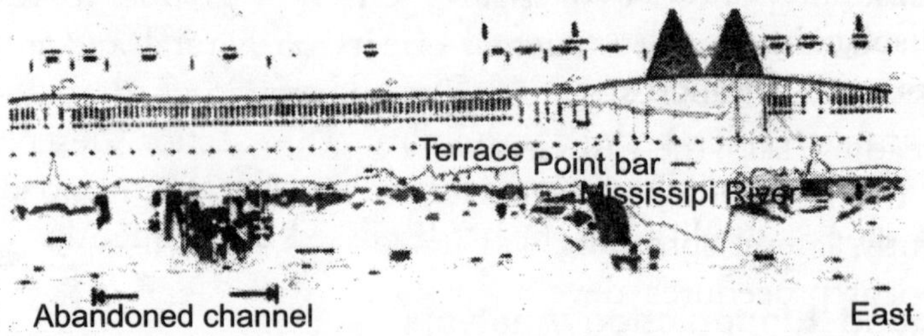

Fig. 14.7 Potential zones of liquefaction along bridge alignment (shown as shaded regions) (Courtesy: https://netfiles.uiuc.edu)

14.7.2 Liquefaction Induced Settlements

In reaches that experience liquefaction, post-earthquake settlements occur. This happens because shaking-induced excess pore water pressure dissipates and the liquefied soil resediments and consolidates. Post-earthquake settlement using the Tokimatsu and Seed (1987) SPT based procedure was computed.

The magnitude of liquefaction-induced settlements is primarily a function of the liquefied layer thickness and the severity of liquefaction (i.e. FS). As the liquefied layer thickness increases or the FS decreases, liquefaction-induced settlement increases. Fig. 14.8 presents the settlement calculations graphically. These computed settlements are very approximate and could occur as either total or differential settlements. Liquefaction-induced settlements of overlying cohesive or nonliquefied soils typically will result in downdrag on deep foundation elements. Furthermore, in areas of potential lateral spreading, e.g., the current point bar, vertical displacements could be much larger as the result of the lateral movement of liquefied materials.

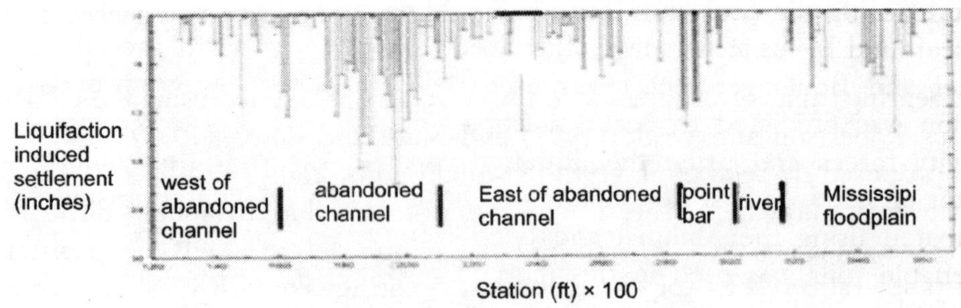

Fig. 14.8 Estimated liquefaction induced settlements (Courtesy: https://netfiles.uiuc.edu)

14.7.3 Lateral Spreading

Lateral spreading occurs in mildly-sloping ground. It happens when the combined downslope static and earthquake-induced inertial forces exceed the undrained resistance of the soil. Since the inertial forces are required to exceed the resistance of the soil, lateral movement usually ceases when shaking ends. Modeling of lateral spreading is complicated by the complex non-linear behavior of soils under seismic loading. As a result, the state-of-the-practice for lateral spreading analysis is based on empirical relationships to predict lateral displacements. These empirical relations account for various soil properties, seismic parameters, and geometric conditions.

Lateral spreading at the Arkansas and Mississippi river banks were evaluated. The remainder of the site (with the exception of the embankments and slopes) generally is level. Consequently, lateral spreading was judged not to be a concern for the remainder of the site. Empirical procedures developed by Bartlett and Youd (1995) as well as by Rauch et al. (2000) was used to estimate magnitudes of lateral displacement. The Bartlett and Youd (1995) as well as Rauch and Martin (2000) methods provide estimates of lateral displacement for both free-face and sloping ground conditions. Sloping ground model for the Arkansas bank and the free-face model for the Mississippi bank was considered in the study.

Using an average slope of 4%, a lateral displacement ranging from 3 to 13 feet for the Arkansas river bank was estimated. The zones of soil subject to lateral spread [i.e., $(N_1)_{60} < 15$] at the Mississippi river bank were found to be discontinuous. As a result it was judged that lateral spreading of this bank was not likely. Based on the study, it was concluded that with a permanent concrete revetment placed on the bank, the bank geometry and potential for lateral spreading was unlikely to change significantly over the structure's lifetime.

14.7.4 Seismic Slope Stability

Seismic slope stability analyses for the approach embankments, the Arkansas and Mississippi USCOE levees, and the natural terrace on the Arkansas side of the river was conducted. The seismic slope stability analysis first considered whether or not the foundation soils underlying the slope are contractive (i.e., susceptible to liquefaction flow failure) using the procedure proposed by Olson and Stark (2003). If the foundation soils were found to be contractive, the Seed and Harder (1990) procedure (as updated by Harder and Boulanger 1997) was used to evaluate the triggering of liquefaction. If liquefaction was predicted to occur, a slope stability analysis was conducted considering only gravity forces and using the liquefied shear strength. If flow failure was unlikely (i.e., factor of safety against flow failure > 1), maximum horizontal seismic displacements was estimated using the Makdisi and Seed (1978) method. Liquefied shear strength in the liquefiable soils was used in this method. The horizontal deformation was capped to that required to mobilize the liquefied shear strength based on a limiting shear strain of 50% (Byrne 1991).

At the approach embankments, liquefaction of thin, but continuous, sandy foundation layers was predicted. Design charts relating embankment height, fill slope angle, and fill material (sand versus clay) to factor of safety for slope stability were developed. Fig. 14.9 presents an example stability analysis for the Arkansas approach embankment. In addition, potential seismic lateral displacements for each trial embankment was also considered. Based on a cost analysis comparing embankment height, bridge costs, predicted lateral displacements, and soil improvement costs (to limit lateral displacement), the design team decided to minimize embankment heights and decided not to use soil improvement at the approach embankments. The final embankment configurations were taken as 2H:1V side slopes and heights of 17.5 ft in Arkansas and 12 ft in Mississippi.

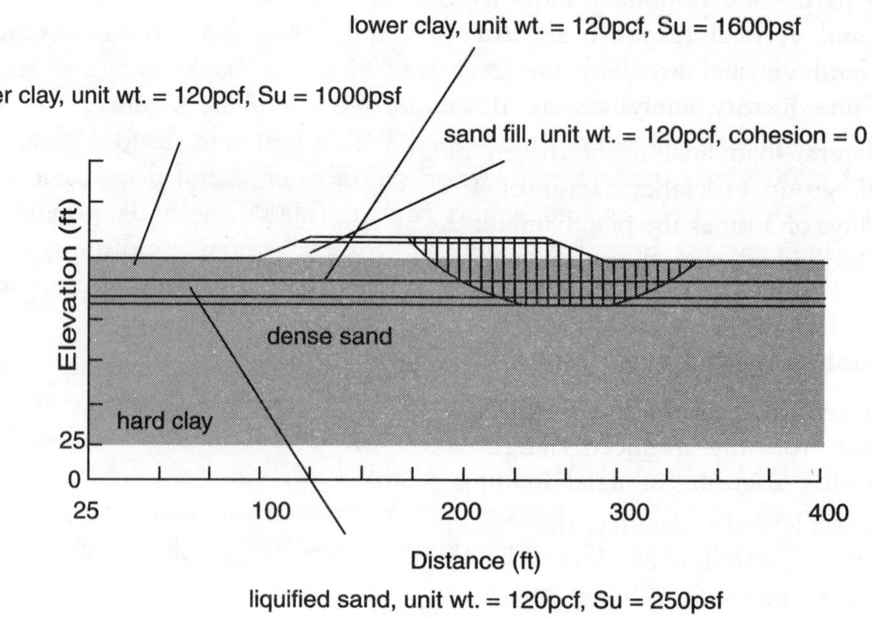

Fig. 14.9 Example seismic slope stability analysis
(Courtesy: https://netfiles.uiuc.edu)

At the Arkansas USCOE levee, it was concluded that liquefaction was likely in a sandy layer about 25 to 30 feet below grade. Due to this depth of the layer, a seismic lateral displacement of less than 4 inches was predicted. Thus, no special precautions were taken at the Arkansas levee. At the Mississippi levee however, lateral displacements on the order of 3 feet was predicted. This magnitude of prediction was due to sliding on a near-surface liquefiable layer. To account for potential liquefaction-induced displacements, it was decided to move the bridge piers outside the zone of displacement. At the natural terrace, a loose sandy silt deposit is present. Extensive liquefaction occurence was predicted in this sandy silt resulting in a liquefaction flow failure. The toe of the terrace is located near a bridge pier; therefore, soil improvement at this location was considered. It has been discussed subsequently.

14.8 FOUNDATION DESIGN RECOMMENDATIONS

Foundation design recommendations for three conditions: (1) no liquefaction; (2) level ground liquefaction; and (3) liquefaction with lateral ground displacement were provided.

14.8.1 Reaches with No Liquefaction

Ground shaking poses a significant design concern for the bridge foundations. Reaches with no liquefaction are going to experience significant ground motion amplification. For example in the abandoned channel, ground response analyses indicated that at a period of 1 second, horizontal motions will be amplified about 5 times that of the moderately hard rock boundary motions. For structural multi-modal analysis, smoothed horizontal and vertical response spectra were provided. In addition, spectrally-matched horizontal and vertical acceleration time histories at the top of foundation level for structural time history analysis were developed.

For lateral load analysis of driven pile and drilled shaft groups in reaches with no liquefaction, group efficiency factor of 1.0 and 0.5, respectively was recommended., for typical spacings of 3 times the pile diameter. An average value for the group was recommended in lieu of individual rows because seismic shaking occurs randomly rather than in one direction.

14.8.2 Reaches with Level Ground Liquefaction

Because liquefaction likely will occur during shaking, the foundation analysis must account for the reduced lateral stiffness of the liquefied soil during shaking and changes in side friction for axial loading.

To assess lateral resistance, the soft clay p-y model (Matlok 1970) was recommended. Furthermore, liquefied shear strengths that increase with depth based on a liquefied strength ratio approach (Olson and Stark 2002) was also developed. This approach yielded results consistent with the approach recommended by Po Lam et al. (1998) in liquefied soil. In areas of marginal liquefaction, installing displacement piles with spacing-to-diameter (s/d) ratios of about 3 is likely. This will densify the soil sufficiently to preclude liquefaction. In zones of more severe liquefaction, it was anticipated that pile installation would not preclude liquefaction. Therefore, all piles should use reduced p-y values. For post-earthquake analyses, all the soils will provide lateral resistance to pile movement. Materials above the bottoms of the pile caps generally were non-liquefiable and it was recommended that passive resistance behind the cap be incorporated in design.

During liquefaction, it was recommended that side friction from the liquefied soils and cohesive soils overlying the liquefied soils be discounted. Following the earthquake, shaking-induced excess porewater pressures will begin to dissipate in the liquefied soils leading to settlement of the overlying deposits. This settlement will cause downdrag on deep foundation elements. The actual zones of soil that liquefy will however not cause downdrag on deep foundation elements.

14.8.3 Reaches with Lateral Soil Displacement

Based on the study, liquefaction-induced lateral soil displacement was anticipated at the natural terrace, the Arkansas riverbank, and the Mississippi levee. Lateral soil displacement induces lateral forces on foundation elements. There is large uncertainty in estimating liquefaction-induced lateral forces on foundations. Recent studies from Mageau and Stauffer (1998) based on case history back-analysis suggest that liquefied soil (with strengths similar to that at the project site) exerts a force of 1500 psf on large diameter (±20-ft) drilled shaft foundations. Arching effects between closely spaced piles are unknown.

Lateral soil pressure of 1500 psf on the projected area of the foundation and piles or shafts was recommended. If this soil pressure were present, it was suggested that non-liquefied soils above the zone of liquefaction will apply a passive pressure on an area that is 1.5 times wider than the projected area of the foundation. Based on empirical studies it has been suggested that this passive pressure should be calculated using Rankine theory. Due to the large lateral forces and uncertainties in estimating these forces, it is recommended that liquefaction mitigation be utilized where practical rather than designing the foundations to resist liquefaction-induced forces.

14.9 LIQUEFACTION MITIGATION CONCEPTS

Options to mitigate liquefaction at the Arkansas terrace and the Arkansas riverbank were explored. At the Mississippi levee, to mitigate earthquake effects, the design team moved the nearest bridge pier outside of the likely zone of lateral displacement.

14.9.1 Arkansas Terrace

It was anticipated that liquefaction of the sandy silts underlying the terrace will trigger a flow failure of the terrace slope and threaten a bridge pier. Several options to mitigate liquefaction at this location were considered. They included: (1) Flattening the slope and installing a rock key trench; (2) Deep dynamic compaction; and (3) Deep soil mixing. In all cases, flattening the slope would reduce the likely failure mode from flow failure to lateral spreading. Consequently, it was highly recommended. Detailed cost comparisons of these options was not done. However, preliminary costs for options (1) and (2) were carried out.

It was assumed that a zone of soil should be improved for both options. Sheetpiling may be driven (and later extracted) to allow vertical excavation. Assuming typical excavation costs of \$6/yd^3 and shot rock replacement costs of \$25/yd^3, the overall cost of this improvement option was about \$550,000. The unit cost of deep dynamic compaction (DDC) is about \$1.50/ft^2, resulting in an improvement cost of about \$120,000. However because of the high silt content of the soils underlying the terrace, it was recommended that a field test be performed to verify the effectiveness of DDC. In addition, it was also suggested that a gravel pad be placed over the entire area to be densified prior to production of DDC work to improve its effectiveness.

14.9.2 Arkansas Riverbank

It was anticipated that the Arkansas riverbank would experience lateral spreading of the order of 3 to 13 feet. The depth of lateral spreading was expected to be between 15 and 50 feet. As a result, flattening the slope and providing scour protection would reduce the overall magnitude of displacement but it would not prevent lateral movements and the resulting forces on the foundations. Because of the thickness of liquefiable soils and the difficulty in accurately estimating lateral spreading forces, it was strongly recommended that soil improvement and subsequent bank protection in this area should be carried out. Available mitigation options at this location included: (1)vibratory compaction; (2) explosive compaction; (3) deep soil mixing; and (4)jet or compaction grouting. In all cases, erosion protection for the riverbank was recommended so that the improved soil is not removed by scour.

At four foundation units, it was estimated that the depth of lateral spreading would approach 50 feet. In this reach, it was anticipated that improvement by vibrocompaction would be most economical. It was estimated that 3-foot diameter vibro-compaction columns on an 8-foot spacing to a depth of 55 feet (i.e., 5-foot penetration into dense sand) would be sufficient in this area to resist the lateral spreading forces and not transmit lateral spreading forces to the caisson. The required zone of improved soil is shown schematically in (Fig. 14.10). The improvement cost for each caisson is about \$1.1 million assuming a unit cost of \$10/yd^3. (This estimated price does not includes a premium for barge work.). At two foundation units, it was estimated that the depth of lateral spreading would decrease to about 20 feet. In this reach, making the same assumptions as above but decreasing the size of the improved zone, the improvement cost was about \$200,000 per caisson.

14.10 CONCLUSIONS

The Great River Bridge, still under design (in 2003), has proven to be a very challenging project due to the stringent seismic design criteria. Predicted magnitudes of ground shaking, liquefaction, and liquefaction-induced lateral soil displacements have forced the design team to evaluate more detailed analysis techniques and liquefaction mitigation concepts. As design progresses, these options will be refined and incorporated into the bridge design to achieve a constructible and cost-effective solution.

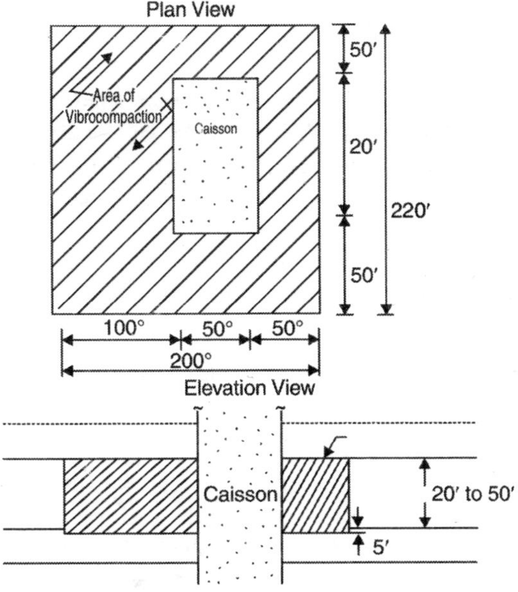

Fig. 14.10 Preliminary liquefaction mitigation concept for the Arkansas riverbank (Courtesy: https://netfiles.uiuc.edu)

REFERENCES

Abrahamson, N.A. (1993). Personal communication to Dr. Joseph Sun.

Acharrya, S.K. 1999. Natural Hazards: Earthquakes and Landslides in India during the nineties and mitigation strategy. Proc. Sem. on Earthquakes & landslides: Natural disaster mitigation,Calcutta, pp. 5–13.

Ahmed, K. 1998. Special report of earthquakes. Down to Earth, December 31 issue, pp. 22–23.

Ambraseys, N. and D. Jackson, (2003). A note on early earthquakes in northern India and southern Tibet, Current Science, **84**(4), 571–582.

Ambraseys, N. and R. Bilham (2003**a**). Earthquakes in Afghanistan, *Seism. Rev. Lett.,* **74**(2), 107–123.

Andrus, R. D. and Stokoe, K.H., II (1997). "Liquefaction resistance based on shear wave velocity". *Proc., NCEER Workshop on Evaluation of Liquefaction Resistance of Soils, Nat. Ctr. for Earthquake Eng.* Res., State Univ. of New York at Buffalo, 89–128.

Arango, I. (1996). "Magnitude scaling factors for soil liquefaction evaluations". *J. Geotech. and Geoenvir. Eng., ASCE.* Vol. **126**(11): 929–936.

Atkinson, G., B. Bakun, P. Bodin, D. Boore, C. Cramer, A. Frankel, P. Gasperini, J. Gomberg, T. Hanks, B. Herrmann, S. Hough, A. Johnston, S. Kenner, C. Langston, M. Linker, P. Mayne, M. Petersen, C. Powell, W. Prescott, E. Schweig, P. Segall, S. Stein, B. Stuart, M. Tuttle, and R. Van Arsdale (2000). "Reassessing New Madrid." *Proc., 2000 New Madrid Source Workshop.*

Australian Building Codes Board, 1996. Building Code of Australia. CCH Australia for the ABCB. Canberra.

Bapat, A. 1996. Creation of awareness about earthquakes—Case Histories. Prc. Int. Conf. on Disaster & Mitigation, Madras, Vol. 1, pp. A1 1–3.

Bartlett, S.F. and T.L. Youd (1995). "Empirical prediction of liquefaction-induced lateral spread." *J. Geot. Eng.*—ASCE, Vol. 121, No. 4, pp. 316–329.

Bendick, R. and R. Bilham (1999). Search for buckling of the southwest Indian coast related to Himalayan collision. In Himalaya and Tibet: mountain roots to mountain tops.

Bhagwan, N.G. and Sreenath, H.G. 1996. A seismic vulnerability reduction strategy for low rise building construction. Int. Conf. on Disaster & Mitigation, Madras, Vol. 1, pp. A311–16.

Bilham, R. (1994). The 1737 Calcutta Earthquake and Cyclone Evaluated, *Bull. Seism. Soc. Amer.* **84**(5), 1650–1657.

Bilham, R., F. Blume, R. Bendick and V. K. Gaur (1998) Geodetic constraints on theTranslation and Deformation of India: implications for future great Himalayan earthquakes, *Current Science*, **74**(3), 213–229.

Bilham, R., R. Bendick and K. Wallace (2003). Flexure of the Indian Plate and Intraplate Earthquakes, *Proc. Indian Acad. Sci. (Earth Planet Sci.)*,**112**(3), 1–14.

Byrne, P.M. (1991). "A model for predicting liquefaction induced displacements." *Proc., 2nd Intl. Conf. Recent Advances in Geot. Earthquake Eng. and Soil Dyn., St. Louis,* MO, Vol. II, pp. 1027–1035.

Chester *et al.*, Journal of Geophysical Research, 1993.

Cooling, T.L., S.M. Olson and S.T. Hague (2003). "Site response and spatial incoherency analysis for the Great River Bridge." Manuscript submitted to 4th Natl. Seismic Conf. and Workshop on Bridges and Highways, February 9–11, 2004, Memphis, TN.

Cunny, R.W. and Sloan, R.C. (1961). "Dynamic loading machine and results of preliminary small-scale footing tests" A.S.T.M. *Symposium on Soil Dynamics*, Special Technical Publications, No. 3–5 pp. 65–77.

Cushing, E.M., E.H. Boswell and R.L. Hosman (1964)."General geology of the Mississippi Embayment." U.S. Geological Survey Professional Paper 448-B, Washington, D.C.

Das Gupta, S.K., 1999. Seismic hazard assessment: local geological effects of strong ground motion. Proc. Sem. on Earthquakes and Landslides: Natural disaster mitigation, Calcutta, pp. 44–51.

Das, T. and Sarmah, S.K., 1996. Earthquake expectancy in northeast India. Int. Conf. on Disaster & Mitigation, Madras, Vol. 1, pp. A1. 14–18.

Day, R. W. (2002). Geotechnical Engineering Handbook, *Mcgraw-Hill Handbooks.*

Department of Science & Technology, Govt. of India (1999). Earthquake Researches in India, p.13.

Ed. A. Macfarlane, R. Sorkhabi and J. Quade. *Geological Society of America Special Paper* **328**. 313–323.

Ervin, C.P. and L.D. McGinnis (1975). "Reelfoot Rift: Reactivated precursor to the Mississippi Embayment." *Geol. Soc. of America Bull.* **86**, pp. 1287–1295.

Frankel, A.D., C. Mueller, T. Barnhard, D. Perkins, E.V. Leyendecker, N. Dickman, S. *Hanson* and M. Hopper (1996). "National seismic hazards maps: Documentation." *U.S. Geological Survey Open File Report,* 96–532.

Gupta, H.K., 1999. Big quakes more a norm than exception. *Times of India*, September 19,1999, p. 12.

Harder, L.F., Jr. and R. Boulanger (1997). "Application of K. and 1K correction factors." *Proc., NCEER Workshop on Evaluation of Liquefaction Resistance of Soils*, National Center for Earthquake Engineering Research Technical Report NCEER-97-0022, pp. 167–190.

Idriss, I. M. (1999). "An update of the Seed-Idriss simplified procedure for evaluating liquefaction potential". *Proc., TRB Workshop on New Approaches to Liquefaction* Analysis, FHWA-RD-99-165. Federal Highway Administration, Washington, D.C.

Ishihara, K. and Yoshimine, M. (1992). "Evaluation of settlements in sand deposits following liquefaction during earthquakes". *Soils and Foundations*, Vol. 32(1): 173–188.

Iyengar, R. N., 1994. Earthquake History of South India, *The Hindu*, Jan. 23.

Katsumi, Maraya, M.M. and Miteuru, T. (1988). "Analysis of gravel drain against liquefaction and its application to design" *IXth WCEE, Tokyo,* Vol. III, pp. 249–254.

Kayal, J.K. 1998. Seismicity of northeast India and surroundings—development over the past 100 years. Jour. Geophysics, XIX (1), pp. 9–34.

Khan, K. (1874). Muntakhab-ul Lubab, M.H., Bibl. India Series, Calcutta.

Khatri, K.N., 1999. Probabilities of occurrence of great earthquakes in the Himalayas. Earth & Planetary Sciences. **108** (2), pp. 87–92.

Kramer, S.L. (1996). "Geotechnical Earthquake Engineering". *Prentice-Hall.*

Kuwabara, F. and Yoshumi, Y. (1973). "Effect of subsurface liquefaction on strength of surface soil" *ASCE, JGE,* Vol. 19, No. 2.

Lee, K.L. and Albaisa, A. (1974), "Earthquake induced settlement in saturated sands". *Journal of the Soil Mechanics and Foundations Division, ASCE,* Vol. 100, No. GT4.

Lew, M. (1984). "Risk and mitigation of liquefaction hazard", *Proc. VIIIth WCEE,* Vol. 1, pp. 183–190.

Lilhanand, K. and W.S. Tseng (1988). "Development and application of realistic earthquake time histories compatible with multiple damping response spectra," *9th World Conf. Earth. Eng.,* Vol. II, pp. 819–824.

Lyman, A.R.N. (1942). "Compaction of cohesionless foundation soil by explosive," *ASCE Trans.*, Vol. 107.

Mageau, D. and S. Stauffer (1998). "Lateral spread earth pressures on drilled shafts."

Proc., Geot. Earthquake Eng. and Soil Dyn. III, ASCE Geot. Spcl. Pub. No. 75 (P. Dakoulas, M. Yegian, and R.D. Holtz, eds.), Vol. 2, pp. 1344–1355.

Makdisi, F.I. and H.B. Seed (1978). "Simplified procedure for estimating dam and embankment earthquake-induced deformations." *J. Geot. Engrg. Div.,* ASCE, Vol. 104, No GT7, pp. 849–867.

Matlock, H. (1970). "Correlations for design of laterally loaded piles in soft clay." *Proc., 2nd Annual Offshore Technology Conf.,* Houston, Texas, Vol. 1, pp. 577–594.

Meyerhof, G.G. (1951). "The ultimate bearing capacity of foundations," *Geotechnique,* Vol. 2, No. 4, pp. 301–331.

Meyerhof, G.G. (1953). "The bearing capacity of footings under eccentric and inclined loads", *Proc. Third International Conf. Soil Mech. Foun. Engg.*, Zurich, Vol. 1, pp. 440–445.

Nandi, D.R. 1999. Earthquake, earthquake hazards and hazards mitigation with special reference to India. Proc. Sem. on Earthquake & Landslides: Natural disaster mitigation, Calcutta, pp. 15–20.

New Zealand Government Print, 1992. Regulations to the Building Act, Wellington Oldham, T. (1883). A Catalogue of Indian earthquakes, *Mem. Geol. Surv. India,* 19, 163 215,Geol. Surv. India, Calcutta.

Nuttli, O.W. (1973). "The Mississippi Valley earthquakes of 1811 and 1812: Intensities, ground motion and magnitudes." *Bull. Seism. Soc. Am.,* Vol. 63, No. 1, pp. 227–248.

Nuttli, O.W. (1982). "Damaging earthquakes of the central Mississippi Valley," in *Investigations of the New Madrid, Missouri Earthquake Regions,* (F.A. McKeown and L.C. Pakiser, eds.) *U.S. Geological Survey Professional Paper* 1236, pp. 15–20.

Ohsaki, Y. (1970). "Effects of sand compaction on liquefaction during Tokachioki earthquake". Soils and Foundations, Vol. **10**(2), 112–128.

Olson, S.M. and T.D. Stark (2002). "Liquefied strength ratio from liquefaction flow failure case histories." Canadian Geot. J., Vol. 39, pp. 629–647.

Olson, S.M. and T.D. Stark (2003). "Yield strength ratio and liquefaction analysis of slopes and embankments." *J. Geot. and Geoenv. Eng.,* ASCE, Vol. 129, No. 8, pp. 727–737.

Paul, J., Burgmann, R. Gaur, V. K. Bilham, R. Larson, K. M. Ananda, M. B. Jade, S. Mukal, M. Anupama, T. S. Satyal, G., Kumar, D. 2001. The motion and active deformation of India. Geophys. Res. Lett. **28**(4), 647–651, 2001.

Paulay, T. 1997. A Review of Code Provisions for Torsional Seismic Effects in Buildings. New Zealand National Society for Earthquake Engineering Bulletin. Wellington, Vol. **30** (3)pp. 252–264.

Po Lam, I., M. Kapuskar and D. Chaudhuri (1998). "Modeling of pile footings and drilled shafts for seismic design." *Tech. Rpt. MCEER-98-0018,* Multi-Disciplinary Center for Earthquake Engineering Research, University of Buffalc, New York.

Rajendran, C.P. (2000). Using geological data for earthquake studies: A perspective from peninsula India, *Current Science*, **79**(9), 1251–1258.

Rajendran, C.P. and Rajendran, K. (2002). Historical Constraints on Previous Seismic Activity and Morphologic Changes near the Source Zone of the 1819 Rann of Kachchh Earthquake: Further Light on the Penultimate Event., *Seism Res. Lett.*, **73**(4), 470–479.

Rajendran, C. P., K. Rajendran, K. H. Vora and A. S. Gaur (2003). The odds of a seismic source near Dwarka, NW Gujarat: An evaluation based on proxies, *Current Science*, **84**, 695–701.

Rao Ramachandra, Publication No. 1, Central Board of Geophysics, Gov. of India, 1953.

Rauch, A.F. and J.R. Martin II (2000). "EPOLLS model for predicting average displacements on lateral spreads." *J. Geot. and Geoenv. Eng.,* ASCE, Vol. 126, No. 4, pp. 360–371.

Ray, 1953, Isoseismals for the great Assam Earthquake of Aug. 15, 1950, 35–37, *in* A compilation of papers on the Assam Earthquake of August 15, 1950. ed. M. B.

Robertson, P. K. and Wride (Fear), C.E. (1997). "Cyclic liquefaction and its evaluation based on the SPT and CPT." *Proc., NCEER Workshop on Evaluation of Liquefaction Resistance of Soils,* (T.L. Youd and I.M. Idriss, eds.) National Center for Earthquake Engineering Research Technical Report NCEER-97-0022.

Romero, S.M. and G.J. Rix (2001). "Ground motion amplification of soils in the upper Mississippi Embayment." *Report No. GIT-CEE/GEO-01-1,* Georgia Institute of Technology, Atlanta, Georgia.

Saucier, R.T. (1994). "Geomorphology and Quaternary geologic history of the lower Mississippi valley." *U.S. Army Corps of Engineers Waterways Experiment Station*, Vicksburg, Mississippi, prepared for Mississippi River Commission, Vicksburg, Mississippi.

Sarmah, S.K. 1999. The probability of occurrence of a high magnitude earthquake on Northeastern India. Jour. of Geophysics, Vol. XX(3), pp. 129–135.

Schnabel, P. B. and Seed, H. B. Acceleration in rock for earthquakes in Western United States, Bulletin of the Seismological Society of America, Vol. 63, No. 2, 1973.

Schnabel, P. B., J. Lysmer and H.B. Seed (1972). "SHAKE: A computer program for earthquake response analysis of horizontally layered sites." *Report No. UCB/EERC* 72/12, Earthquake Engineering Research Center, University of California, Berkeley.

Seeber, L. and V. Gornitz (1983). River profiles along the Himalayan arc as indicators of active tectonics, *Tectonophysics*, **92**, 335–367.

Seed, R.B. and L.F. Harder Jr. (1990). "SPT-based analysis of cyclic pore pressure generation and undrained residual strength." *Proc. H. Bolton Seed Memorial Symp.,* Bi-Tech Publishing Ltd., Vol. 2, pp. 351–376.

Seed, H.B. and I.M. Idriss (1971). "Simplified procedure for evaluating soil liquefaction potential." *J. Soil Mech. and Fndtn. Eng. Div.,* ASCE, Vol. 97, No. SM9, pp. 1249–1273.

Seed, H.B. and Lee, K.L. (1966). "Liquefaction of saturated sands during cyclic loading", *ASCE, JGE*, Vol. **92**, No. SM 6, pp. 105–134.

Seed, H. B. and Idriss, I. M. (1970) Soil moduli and damping factor for dynamic response analyses, Report EERC 70–10, Earthquake Engineering Research Center, University of California, Berkeley.

Seed, H.B. and Whitman, R.V. (1970). "Design of earth retaining structures for dynamic loads", *Proceedings, ASCE speciality conference on lateral stresses in the ground and design of earth retaining structures*, ASCE, pp. 103–147.

Seed H. B. and Idriss, I.M. (1971). "Simplified procedure for evaluating soil liquefaction potential", *Journal of Soil mechanics and Foundations Division*, ASCE 97, SM9, pp. 1249–1273.

Seed, H.B., Arango, I. and Chan, C.K. (1975). "Evaluation of soil liquefaction potential during earthquakes," *Report on EERC*, 75–28, Earthquake engineering research center, University of California, Berkeley.

Seed, H. B., Murnaka, R., Lysmer, J. and Idris, I. Relationship between maximum acceleration, maximum velocity, distance from source and local site conditions for moderately strong earthquake, EERC 75–17, University of California, Berkeley, 1975.

Seed, H.B., Tokimatsu, K. and Harder, L. (1984). "The influence of SPT procedures in evaluating soil liquefaction resistance". *Report No. UCB/EERC-84-15,* Earthquake Engg. Res. Ctr., Univ. of California, Berkeley, Calif.

Seed, H.B., Tokimatsu, K. Harder, L.F. and Chung, R. (1985). "Influence of SPT procedures in soil liquefaction resistance evaluations". *J. of Geotechnical Engineering, ACSE*, Vol. **111**, GT 12, pp. 1425–1445.

Sella, G. F., T. H. Dixon, and A. Mao, (2002). REVEL: A model for recent plate velocities from space geodesy, *J. Geophys. Res.,* **107**, 10.1029/2000JB000033.

Skempton, A. W. (1986). "Standard Penetration Test Procedures and the Effects in Sands of Overburden Pressure, Relative density, Particle size, Aging and Overconsolidation", *Geotechnique*, Vol. **36**, no. 3, pp. 425–447.

Somerville, P.G., J.P. McLaren, C.K. Saikia and D.V. Helmberger (1990). "The November 25, 1988, Saguenay Quebec earthquake: source parameters and the attenuation of strong ground motion." *Bull. Seism. Soc. Am.,* Vol. 80, pp. 1118–1143.

Somerville, P., N. Collins, N. Abrahamson, R. Graves and C. Saikia (2001). "Ground motion attenuation relations for the central and eastern United States." *U.S. Geological Survey Final Report Award No. 99HQGR0098.*

Stark, T.D. and S.M. Olson (1995). "Liquefaction resistance using CPT and field case histories." *J. Geot. Engrg.,* ASCE, Vol. **121**, No. 12, pp. 856–869.

Struck, D. 1999. Tokyo prepares for an overdue disaster, *Times of India*, October 11, 1999, p.12.

Sukhija, B. S., M. N. Rao, D. V. Reddy, P. Nagabshanam, S. Hussain, R. K. Chadha and H.K. Gupta (1999). Timing and return of major paleoseismic events in the Shillong Plateau, India, *Tectonophysics,* **308**, 53–65.

Susumu, I., Koizimi, K., Node S. and Ysuchia, H. (1988). "Large scale model tests and analysis of gravel drains," *IXth WCEE, Tokyo,* Vol. **III**, pp. 261–266.

Swami Saran (1999). "Soil dynamics and machine foundations", *Galgotia Publications,* New Delhi.

Tandon, A. N., The Very Great Earthquake of Aug. 15, 1950, 80–89, *in* A compilation of papers on the Assam Earthquake of August 15, 1950. ed. M. B. Ramachandra Rao, Publication No.1, Central Board of Geophysics.

Tatsuoka, F., Sasaki, T. and Yamada, S. (1984). "Settlement in saturated sand induced by cyclic undrained simple shear". *J. of Geotechnical Engineering, ASCE.*

Tokimatsu, K. and Seed, H.B. (1987). "Evaluation of settlements in sand due to earthquake shaking". *J. of Geotechnical Engineering, ASCE.* Vol. **113**(8):, 861–878.

Terzaghi, K. (1943). "Theoretical soil mechanics", *John Wiley and Sons*, New York.

Terzaghi, K. and Peck, R. B. (1967). "Soil mechanics in engineering practice" *Ist Edition, John Wiley and Sons*, New York.

Triandafilidis, G. E. (1965). "Dynamic response of continuous footings supported on cohesive soils" *Proc. Sixth Int. Conf. Soil Mech. Found. Engin.,* Montreal, Vol. **2**, pp. 205–208.

Vesic, A.S. (1973). "Analysis of ultimate loads of shallow foundation", *Journal of Soil mechanics and Foundations division,* ASCE, Vol. **99**, SM1, pp. 45–73.

Wallace, W. L. (1961), "Displacement of long footings by dynamic loads" *ASCE Journal of Soil mechanics and Foundation division*, **87**, SM5, pp. 45–68.

Wang, Qi, Pei-Zhen Zhang, J. T. Freymueller, R. Bilham, K. M. Larson, XiÕan Lai, X. You, Z. Niu, J. Wu, Y. Li, J. Liu, Z. Yang, Q. Chen, Present Day Crustal Deformation in China constrained by Global Positioning Measurements, *Science*, **294**, 574–577, 2001.

Wesnousky, S. G., S. Kumar, R. Mohindra, and V.C. Thakur (1999). Holocene slip rate of the Himalaya Frontal Thrust of India-Observations near Dehra Dun, *Tectonophysics*, **18**, 967–976.

Wheeler, R.L. and D.M. Perkins (2000). "Research, methodology, and applications of probabilistic seismic-hazard mapping of the central and eastern United States-Minutes of a workshop on June 13–14, 2000, at Saint Louis University." *U.S. Geological Survey Open File Report 00-0390.*

White, C. R. (1964). "Static and dynamic plate bearing tests on dry sand without overburden", Report R. 277, U.S. Naval Civil Engineering Laboratory.

Wright, D. (1877). History of Nepal, 1966 reprint: Calcutta, Ranjan Gupta, 271.

Yoshimi, Y. (1967). "Experimental study of liquefaction of saturated sands" *Soil Found.* (*Tokyo*),Vol. 7, No. 2, pp. 20–32.

Youd, T.L. and Idriss, I.M. (1997). "Summary report." Proc., NCEER Workshop on Evaluation of Liquefaction Resistance of Soils, (T.L. Youd and I.M. Idriss, eds.) National Center for Earthquake Engineering Research Technical Report NCEER-97–0022.

Youd, T. L. and Noble, S. K. (1997). "Magnitude scaling factors". *Proc., NCEER Workshop on Evaluation of Liquefaction Resistance of Soils* Nat. Ctr. for Earthquake Engrg. Res., State Univ. of New York at Buffalo, 149–165.

Internet references

<http://cires.colorado.edu>

<http://eqseis.geosc.psu.edu>

<http://gbpihed.nic.in>

<http://seismo.unr.edu>

<http://web.ics.purdue.edu>

<http://www.branz.co.nz>

<http://www.es.ucs.edu>

<http://www.geo.mtu.edu>

<http://www.stvincet.ac.uk>

<http://www.vulcanhammer.net>

INDEX